Towards an Ethical-ecological Assessment of Companies in Nigeria

European University Studies
Europäische Hochschulschriften
Publications Universitaires Européennes

Series V
Economics and Management

Reihe V Série V
Volks- und Betriebswirtschaft
Sciences économiques, gestion d'entreprise

Vol./Bd. 3384

PETER LANG
Frankfurt am Main · Berlin · Bern · Bruxelles · New York · Oxford · Wien

Emmanuel Franklyne Onyemaechi
Ogbunwezeh

Towards an Ethical-ecological Assessment of Companies in Nigeria

An Empirical Inquiry into the Relevance or Otherwise of the Frankfurt-Hohenheim Guidelines for the Ethical Assessment of Companies in the Nigerian Context

A Case of the Nigerian Microfinance Banking Sector

PETER LANG
Internationaler Verlag der Wissenschaften

Bibliographic Information published by the Deutsche Nationalbibliothek
The Deutsche Nationalbibliothek lists this publication in the Deutsche Nationalbibliografie; detailed bibliographic data is available in the internet at http://dnb.d-nb.de.

Zugl.: Frankfurt (Main), Univ., Diss., 2009

D 30
ISSN 0531-7339
ISBN 978-3-631-61696-3

www.peterlang.de

Acknowledgement

In the course of working on this dissertation, I received help and assistance from various individuals and groups, who would forever be dear to my esteem.

I thank my family for their unwavering confidence and faith.

I will remain eternally grateful to the Federal Government and people of the Federal Republic of Germany, who awarded me a scholarship for gifted students, through the Konrad Adenauer Foundation to do this study. Without that financial basis, this work would have not been realized.

I thank Dr. h.c. Berthold Gees, who was my contact with the Konrad Adenauer Foundation, and all other workers and alumni of the foundation. Our exchanges enriched my experiences here in Germany and broadened my views on the implications of life and study in Germany.

I thank my indefatigable Supervisors: Prof. Dr. Johannes Hoffmann, and Prof. Dr. Obiora Ike. These men have been august and of the highest integrity. I thank my colleague Dr. Agi Makil for his assistance.

I thank Mr. Karl Hafen, the Executive Director of the International Society for Human Rights. (IGFM) He is a great human being; who greatly inspired me with his kindness, intelligence and humility.

I thank equally Joanna Asia Opozda who was a great friend and would remain one, for all her support and encouragement even in the darkest moments when all was gloom.

I dedicate this work to my late Siblings Mrs. Obiageli Vivian Edeh (nee Ogbunwezeh) and Mr. Christopher Ogbunwezeh. I miss you both. I will always love you. Rest in peace!

To all those whom I failed to mention here, I forget no one. I can only say thank you for all you are and have been to me.

This second edition of Unusual Diseases with Common Symptoms is dedicated to my family: my deceased parents Li-Chieh and Alice Szema; sister and colleague Clinical Adult and Pediatric Allergist/ Immunologist Katherine Szema, MD, FAAP, FAAAAI; wife, Denise C. Monte, MD, FACS, and children Allison and Austin Szema. I am particularly proud that NYU Gallatin Storytelling undergraduate student Allison Szema is the illustrator for this edition. My late mentors Robert E. Dutton, Jr., MD, Professor of Medicine and Physiology at Albany Medical College and Professor of Biomedical Engineering at Rensselaer Polytechnic Institute and Sami I. Said, MD, Distinguished SUNY Professor provided the training for our team to complete this project.

Preface

By way of introduction and description of what the reader will find in this book, *Unusual Diseases with Common Symptoms*, second edition, illustrates new actual patients with unusual diseases, beginning with a clinical vignette, followed by sections on diagnosis and treatment, key bullet points, and multiple-choice questions with references. Again, also included are photos, diagrams, and tables, as well as illustrations by Allison Szema to emphasize take-home messages. While the level of detail is that which I teach students from Zucker School of Medicine at Hofstra/Northwell and medical students who rotate from schools across the US, as well as the same detail which I use to teach medical students in the Pharmacotherapeutics Course at Renaissance School of Medicine at Stony Brook University, the text is written so that a layperson can be introduced to the subject in an easy-to-read, illustrative manner.

These are my patients with diagnoses such as allergy to galactose alpha 1,3 galactose, Balsam of Peru Gell-Coombs Type IV contact reaction, eosinophilic colitis, postural orthostatic tachycardia syndrome, and common variable immunodeficiency.

Our authors are now medical students and physicians, so I am proud of the team!

If you want to contact us, email our team at threevillageallergyandasthma@gmail.com

Sincerely,

New York, NY, USA Anthony M. Szema

The original version of this book was revised: Preface has been added.

Acknowledgments

I thank my Editor Miranda Finch and Production Editor Vinodh Thomas. As a parent, I am extremely proud and grateful for our illustrator Allison Lee Szema, NYU Gallatin undergraduate.

Contents

Hips Don't Lie

1

Max Lee

M. Lee (✉)
Physical Medicine and Rehabilitation, NYU Langone Health,
New York, NY, USA
e-mail: max.lee@nyulangone.org

A. M. Szema (ed.), *Unusual Diseases with Common Symptoms*,
https://doi.org/10.1007/978-3-031-40280-7_1

Vignette

KZ, who recently celebrated her 50th birthday, had become accustomed to issues with her bones. She grew up an avid sports player, but her progressively worsening scoliosis forced her to quit early. Schroth exercises to align the curvature of her spine proved to be insufficient and an operation was inevitable. KZ underwent surgical correction, and a Harrington rod was screwed into her spine to align and support it. Harrington rods have been the gold standard treatment for severe scoliosis since the 1960s [1], but they carry significant risk, especially if placed in a young child. When these patients hit their adolescent growth spurts, they can outgrow the rigid device. This was the case for KZ, and she eventually required it to be removed. Despite these difficulties, she was able to remain strong and live a relatively normal life—until 1 day her hips suddenly gave way.

It turns out she also had a congenital problem, and her hips had degenerated to the extent that bone grinded against bone! Among orthopedic injuries, hip injuries are notoriously dangerous. A third of patients over the age of 65 may not survive a year after such an injury [2]. A young and otherwise robust individual, KZ fortunately had the odds in her favor. First the surgeons fixed the right side without problems. Then they did the left. For both of her hip replacement procedures, metal-on-metal prostheses were placed. It is named as such because the two components of the joint, the socket of the hip and the head of the femur, are both recreated using metal alloys. The durability is astounding, making it the prosthesis of choice in younger patients who will be relying on it to bear their full weight for years to come. No stranger to metal prosthetics, KZ was determined to make her hip a problem of the past.

Days after the second procedure, she felt pain on the left; it was excruciating, and she felt and heard a pop in her socket. Her leg was dangling out! In desperation, her teenage daughter helped force it back in place. This was subluxation.

She also could not feel anything over the outside of her left thigh. Days went by and her sensation never returned. Instead, it was replaced by pain that worsened every time she tried to walk. To make matters worse, she could not move her leg the way she wanted. Just lifting it to get out of the car became a struggle. What was going on?

In the medical field, this syndrome is well known. We call it meralgia paresthetica (MP), a condition which occurs when the lateral femoral cutaneous (LFC) nerve is compressed for any reason. In the USA where obesity is an epidemic, this is often caused when the weight of an individual's abdomen applies downward pressure on the nerve. Tight clothing like belts and skinny jeans can have a similar effect. However, none of these common causes applied to KZ. The timing was too perfect. She suspected the surgery was to blame.

The surgeon who had placed the prosthesis did not think the symptoms were related to the hip surgery. He commented that he had taken an anterior approach to the hip replacement, which theoretically spares any permanent damage to the muscles or nerves. Yet the surgical approach does cut the fascia of the tensor fascia latae [3]. He told her, "it must be a problem stemming from your scoliotic spine." The surgeon believed that the nerve was being pinched between two vertebrae immediately as it exited the spinal cord and refused to do any more. But KZ pressed on and did independent research on the complications that can arise in a hip surgery.

A year later, her efforts finally paid off. It turned out that the surgeon had neglected KZ's anterior pelvic tilt during the procedure. A pelvis that is tilted forward significantly limits the operative workspace and increases the susceptibility of delicate structures to being injured. To make up for this, he agreed to perform a revisional procedure to correct her pelvic tilt free of charge. After a year of incessant pain, KZ was finally going to find relief. This tightened her up and she no longer felt the instability of subluxation anymore.

Yet the same numbness remained. Her spine had not been the cause of her nerve compression. Frustration was high and hope was low. She went through physician after physician who were unable to pinpoint a cause for her condition. Finally, she found an orthopedic spine surgeon who offered to inject a numbing medication, composed of lidocaine and a steroid, directly into the burning area on her leg.

Instant relief.

The pain was gone. Her sensation was back. She could fully move her leg again! The doctor said this treatment would only last a month or so before she would need another injection. That was fine by KZ. Four days later, the pain and numbness returned in full.

The next step for KZ was to get an MRI of the nerves in her legs to assess the state of her LFC. The short-lived success of the lidocaine injection encouraged us that the nerve was still salvageable, but also that it was being significantly compressed. Whether this was from inflammation, scar tissue, or something else remained difficult to determine.

Hydrodissection was used to split the LFC from the surrounding tissue and give it some room to breath. The procedure was technically a success, but it did not provide the relief that KZ was looking for. The nerve was free from its adhesions, but it had likely suffered permanent damage from being compressed for so long.

KZ was understandably frustrated. Her quest to seek escape from pain had thus far been unfruitful and she was on the verge of giving up. In this moment of darkness, a final opportunity presented itself—nerve regeneration using platelet-rich plasma (PRP). Mostly studied in healing muscle tendons and ligaments, PRP still has a largely experimental role in repairing nerve damage. Yet, we are given hope by promising preliminary data showing PRP to enhance growth of peripheral nerves, likely through its rich supply of growth factors [4].

Modern medicine has come a long way. We can bring people back to life when their hearts have stopped beating. We can create artificial limbs controlled by electrical impulses from the brain. Yet sometimes we are perplexed by seemingly straightforward cases such as KZ's meralgia paresthetica. As we near the limits of what our technologies can offer, only time will tell whether KZ becomes a perpetual victim of her disease or a pioneer in the field of regenerative medicine.

In the end, the problem was solved by straightening the curvature of the spine to eliminate scoliosis and improving the anterior tilt by stretching the hip flexor muscles.

Background/Salient Features

Meralgia paresthetica refers to entrapment of the lateral femoral cutaneous nerve, which subsequently leads to burning, coldness, numbness, and/or tingling over the anterolateral thigh. Nerve damage is a notable complication of hip replacements and numerous case reports have recorded meralgia paresthetica directly caused by such procedures [5, 6]. KZ's case is especially significant because she had multiple risk factors for having a difficult, complicated procedure.

The outcome of a hip replacement is largely dependent on proper placement of the prosthesis. Pelvic tilt is a common condition which can compromise a hip replacement by throwing off its center of gravity. Anterior pelvic tilt can cause impingement between the hip and the femur or dislocate the prosthetic joint anteriorly [7]. The success of the procedure is also affected by spinal deformities. KZ had significant scoliosis, which can cause malalignment of the head of the femur within the acetabular cup [8]. Interestingly, scoliosis can often cause pelvic tilt as a compensatory mechanism for the body to balance itself. The effects of these problems can be minimized by implementing tilt-adjustment to compensate for the rotated pelvis, though such techniques are not standardized and vary by manufacturer [9].

Hydrodissection is a cutting-edge non-surgical procedure which utilizes pressurized saline to separate planes of tissue. This technique is widely used in the treatment of carpal tunnel syndrome [10], and its efficacy in treating other nerve pathologies is under investigation. Only a handful of case studies exist which observe the use of hydrodissection in the treatment of meralgia paresthetica [11, 12]. A physician may choose to use hydrodissection over traditional surgical decompression to better avoid inadvertent injury to the nerves, though no studies evaluating its safety for nerve injections currently exist. It should be noted that this technique carries the added benefit of separating soft tissue adhesions through the creation of new surgical planes [13]. Typically, this procedure is performed by an orthopedic surgeon under ultrasound guidance with the use of local anesthetics.

Platelet-rich plasma involves taking blood from a patient and concentrating the platelet portion into an injectable solution [14]. Using ultrasound guidance, the PRP is injected into a site of injury to accelerate healing. Contrary to steroid injections which aim to reduce inflammation, PRP enhances the inflammatory response by recruiting reparative cells to the area. The human body is good at healing acute injuries, but tends to struggle with those that are chronic. Structures with poor blood supply such as tendons, ligaments, and nerves often fall victim to incomplete repair when damaged. The growth factors and cytokines that are highly concentrated in PRP remind the body that damage exists so that the innate restorative mechanisms can be activated and bring about a natural recovery.

Diagnosis

Meralgia paresthetica is diagnosed clinically, meaning it is determined by a combination of the physical exam and the patient's medical history. It should be suspected when a patient presents with pain, numbness, and/or paresthesia of the skin in the distribution of the upper outer thigh. Risk factors such as obesity and diabetes often clue medical practitioners to the etiology. It is important to evaluate the sensation along the distribution of the lateral femoral cutaneous nerve as part of the physical exam, and the Tinel test is often used for this purpose. It consists of tapping along the course of the nerve, and a test is considered positive if it reproduces the symptoms of pain or paresthesia. Confirmatory testing can be done with a nerve conduction study or ultrasound imaging of the nerve.

Treatment

The two most common causes of MP are obesity and tight clothing. Often, the only treatment required is the cessation of wearing belts and skinny jeans. Other potential causes for compression include but are not limited to: pregnancy, fibroids, ascites, long-distance walking/cycling, and groin trauma. In such cases, greater than 90% of patients see symptomatic improvement through a combination of removing tight clothing, limiting physical activities that extend the hip, applying ice to the affected area, and taking over-the-counter NSAIDs for up to 1 week [15]. Though easier said than done, weight loss is a highly successful treatment modality in patients whose nerve compression stems from central adiposity.

There are certain causes of MP that are less easily fixed. Chronic exposure to lead can lead to permanent nerve damage. A less obvious neurotoxin is glucose, which over time can poison the nerves of diabetic patients. Most pertinent to our case are iatrogenic causes, where medical practitioners inadvertently injure the LFC nerve while performing procedures in the area.

In cases where the damage cannot be reversed, anticonvulsant medications such as carbamazepine, phenytoin, and gabapentin may help relieve the neuropathic pain. Similar to KZ's experience, sensations from the nerve can be blocked through the injection of anesthesia and steroids over the affected area. The last line of treatment is surgery. In patients with intractable pain that has not responded to other modalities, a surgeon can cut the entire nerve out [16]. This removes not only the feeling of pain but also the sensation of anything else. After such a procedure, the patient will have a permanent area of numbness over the distribution of the nerve.

There are newer treatments being tried that have seen variable success. Surgeons have tried to decompress the LFC nerve by cutting parts of the inguinal ligament. KZ underwent hydrodissection, where pressurized water is used to free the nerve from the surrounding tissue. While not considered common practice, there have been cases of success stories where patients saw immediate and permanent relief right after the procedure. Finally there is platelet-rich plasma, a therapy that takes the novel approach of regenerating damaged nerves. With all sources of nerve compression gone, KZ now must rely on her body to heal itself with time. If all goes accordingly, she will be able to escape her pain and move on to the next chapter of her life.

Fun Facts

1. Meralgia paresthetica is a feeling of pain and/or numbness in the upper outer thigh caused by external compression of the lateral femoral cutaneous nerve.
2. Orthopedic procedures of the hip can be complicated by spinal pathology, such as severe scoliosis and anterior pelvic tilt.
3. Hydrodissection is a new technique that can treat refractory MP without resorting to transection of the LFC nerve.

Multiple Choice Questions

1. A 47-year-old man presents with numbness and burning pain over the left upper thigh for the past 3 weeks. His symptoms are worse with prolonged physical activity, especially climbing stairs. BMI is 34 kg/m^2. Which of the following is the best initial treatment for this patient?
   (a) Anticonvulsant
   (b) Neurectomy
   (c) Weight loss
   (d) NSAIDs

2. A 54-year-old woman presents to the emergency department because of numbness and burning pain in her right hand for 8 h. Her symptoms developed while she was working overtime at her office job, but have not subsided with rest. The pain is located over the thenar eminence and palmar surface of the first and second digits. An electrodiagnostic study would provide the best representation of muscle denervation if performed within:
   (a) 2 h
   (b) 2 days
   (c) 2 weeks
   (d) 2 years

3. Each of the following conditions are risk factors for hip surgery complications except for:
   (a) Osteoarthritis
   (b) Rheumatoid arthritis
   (c) Pelvic tilt
   (d) Scoliosis

Answers

1. (c)
   In this patient who is obese and has recent onset of meralgia paresthetica (MP), the source of his nerve compression is most likely due to his weight. Initial management includes reducing pressure over the groin area (avoiding tight clothing) and weight loss, as these are the most common causes of the condition.

(a) Most cases of MP resolve spontaneously and do not require pharmacologic treatment. If symptoms persist for more than a couple months, use of anticonvulsants can be considered to reduce possible neuropathic pain.

(b) Neurectomy of the lateral femoral cutaneous nerve is considered last-line therapy for MP and is not yet indicated in this patient with new-onset symptoms and possibility for spontaneous recovery.

(d) NSAIDs may provide temporary symptomatic relief but would not help with managing the etiology of this patient's MP. Additionally, it is important to treat the underlying cause to avoid complications of prolonged use of NSAIDs.

2. (c)

After a nerve is injured, the degree of damage cannot be fully assessed until Wallerian degeneration is complete. The timeline for this can be anywhere from 1 week to 4 weeks, depending on the length of the injured axon [17].

(a) In severe axonal damage, EMG findings can become abnormal immediately. However, the patient's benign presentation and lack of an abrupt or traumatic cause for her symptoms are not suggestive of significant injury.

(b) See above.

(d) Damaged nerves require growth factors released by viable muscle tissue for recovery. Conversely, muscle depends on electrical activity to function properly. Chronically denervated muscle tissue will degenerate and become fibrotic within 2 years, making this too long of a time period to wait.

3. (a)

OA is one of the most common indications for hip surgery, and almost all patients undergoing a hip replacement procedure will have some degree of OA in their femoroacetabular joint. A hip replacement is the only treatment that provides the removal of degenerative damage in order to restore function.

(b) While rheumatoid arthritis (RA) most commonly affects the small joints of the hands and feet, patients with this disease are more susceptible to infections in all joints of the body relative to the general population. Known as septic arthritis, this condition can be potentially limb-threatening and life-threatening if not caught early. This susceptibility is thought to stem from chronic immunosuppression from glucocorticoids and biologic agents used in the treatment of RA.

(c) and (d) As mentioned previously, any form of spinal deformity can significantly change the pelvic anatomy and increase the number of complications associated with hip surgery. Patients with scoliosis are particularly prone to developing femoroacetabular impingement, where irregular bone growth in the femoroacetabular joint causes the head of the femur and acetabulum to rub against each other during movement, resulting in irritation.

References

1. Clear Scoliosis Institute. Harrington rod: is it right for your scoliosis? 2019. Retrieved from https://clear-institute.org/blog/harrington-rod.
2. Panula J, Pihlajamäki H, Mattila VM, et al. Mortality and cause of death in hip fracture patients aged 65 or older: a population-based study. BMC Musculoskelet Disord. 2011;12:105.
3. Holst DC, Angerame MR, Yang CC. Surgical technique of direct anterior approach for total hip arthroplasty on a standard operating room table. Ann Joint. 2018;3:34. https://doi.org/10.21037/aoj.2018.04.01.
4. Yu W, Wang J, Yin J. Platelet-rich plasma: a promising product for treatment of peripheral nerve regeneration after nerve injury. Int J Neurosci. 2011;121(4):176–80. https://doi.org/10.3109/00207454.2010.544432.
5. Chauhan G, Gupta K, Nayar P. Meralgia paresthetica after total hip arthroplasty in supine position. Saudi J Anaesth. 2013;7(1):105.
6. Pereiro S, Jamart V, Palacio F. Meralgia paresthetica after total hip arthroplasty in supine position: what an anesthesiologist should know. Eur J Anaesthesiol. 2012;29:124–5.
7. Shon WY, Sharma V, Keon OJ, Moon JG, Suh DH. Can pelvic tilting be ignored in total hip arthroplasty? Int J Surg Case Rep. 2014;5(9):633–6.
8. Delsole EM, Vigdorchik JM, Schwarzkopf R, Errico TJ, Buckland AJ. Total hip arthroplasty in the spinal deformity population: does degree of deformity affect hip stability? Spine J. 2017;17(10).
9. Maratt JD, Esposito CI, Mclawhorn AS, Jerabek SA, Padgett DE, Mayman DJ. Pelvic tilt in patients undergoing total hip arthroplasty: when does it matter? J Arthroplast. 2015;30(3):387–91.
10. Malone DG, Clark TB, Wei N. Ultrasound-guided percutaneous injection, hydrodissection, and fenestration for carpal tunnel syndrome: description of a new technique. J Appl Res. 2010;10:116–23.

11. Su YC, Lee CY, Chang CY, Chen LC, Wu YT. Efficacy of nerve hydrodissection with 5% dextrose in chronic meralgia paresthetica. Pain Pract. 2020;20(5):566–7. https://doi.org/10.1111/papr.12870.
12. Mulvaney SW. Ultrasound-guided percutaneous neuroplasty of the lateral femoral cutaneous nerve for the treatment of meralgia paresthetica: a case report and description of a new ultrasound-guided technique. Curr Sports Med Rep. 2011;10(2):99–104. https://doi.org/10.1249/JSR.0b013e3182110096.
13. Bokey EL, Keating JP, Zelas P. Hydrodissection: an easy way to dissect anatomical planes and complex adhesions. ANZ J Surg. 1997;67(9):643–4. https://doi.org/10.1111/j.1445-2197.1997.tb04616.
14. Hospital of Special Surgery. Platelet-Rich Plasma (PRP) injection: how it works. 2020. Retrieved from https://www.hss.edu/condition-list_prp-injections.asp.
15. Williams PH, Trzil KP. Management of meralgia paresthetica. J Neurosurg. 1991;74:76.
16. de Ruiter GC, Wurzer JA, Kloet A. Decision making in the surgical treatment of meralgia paresthetica: neurolysis versus neurectomy. Acta Neurochir. 2012;154:1765.
17. Feinberg J. EMG: myths and facts. HSS J. 2006;2(1):19–21. https://doi.org/10.1007/s11420-005-0124-0.

Lady Windermere's Cough

2

Rachel Radigan and Robert A. Promisloff

R. Radigan (✉)
New York Institute of Technology College of Osteopathic Medicine, Old Westbury, NY, USA

R. A. Promisloff
Medicine, Drexel University College of Medicine, Philadelphia, PA, USA

A. M. Szema (ed.), *Unusual Diseases with Common Symptoms*,
https://doi.org/10.1007/978-3-031-40280-7_2

Elderly patient with severe weight loss and history of severe cavitary disease

Vignette

Agatha Erlynne is a 72-year-old Caucasian lady who presented to the office with a chief complaint of unexplained profound "weight loss" (20 lb). Associated symptoms included: fatigue, weakness,

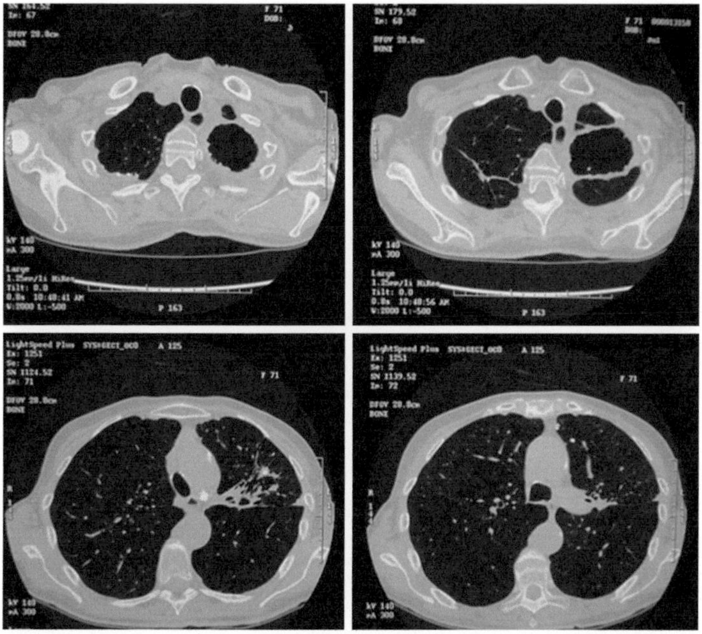

Fig. 2.1 CT chest, 2003. Left upper lobe showing increasing cavitation. Left lower lung showing small nodules

low grade fever, mild shortness of breath, and abdominal pain. Per her primary care physician, extensive workup was done in search of malignancy. Her chest X-ray and CT-scan revealed stable but extensive cavitation in the left upper lobe and small nodules in the left lower lobe (Fig. 2.1).

Agatha has a history of heavy smoking (1–2 packs per day for over 40 years), chronic obstructive pulmonary disease (COPD), an abdominal aortic aneurysm (AAA) repair, and 2 separate symptomatic Nontuberculous Mycobacteria (NTM) infections. Three years earlier she had weight loss, fever, and similar CT findings with progressive cavitation in the left upper lobe and nodular disease in the left lower lobe (Fig. 2.2).

At this time, she was diagnosed with a lung infection due to *Mycobacterium avium Intracellulare* (MAI) via bronchoscopy,

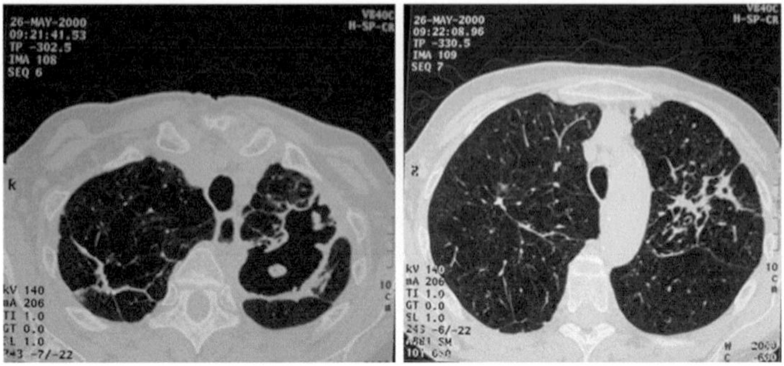

Fig. 2.2 CT chest, 2000. Right upper lung showing severe emphysema. Left upper lung showing large cavitation. Interesting to note *Aspergillus* fungus ball formation in the middle of the cavity in the left lung. Lower left lung showing small nodules (3–5 mm)

which is a long-tubed instrument inserted into the trachea and lungs to visualize the airways and collect tissue samples. Twelve years before that, at the age of 56, also had weight loss, fever, and a cough. A chest X-ray revealed bilateral apical infiltrates and multiple small cavities in the left upper lobe. At this point in time, she was diagnosed with *M. kansasii* by the means of a bronchoscopy and sputum testing. This was a different infection that fell into the same category of atypical *Mycobacterium.*

Could this presentation be *MAI* or *M. kansasii* again, or was it cancer?

Background/Salient Features of the Case

The most common Nontuberculous Mycobacteria (NTM) is *Mycobacterium avium* Complex (MAC). MAC consists of several different species, the most prevalent being *M. avium* and *M. intracellulare*. Diagnosis and treatment are the same for both and differentiation between the two is difficult. MAC and *Mycobacterium avium Intracellulare* (MAI) are terms used interchangeably. The second most common NTM found in patients is *M. kansasii*. It is not uncommon for a patient to be diagnosed with several different species of NTM.

MAI has been commonly found in aerosolized water, piped hot water systems, bathrooms, dust, soil, and in some components of cigarettes such as the tobacco, filter, or paper. MAI is a versatile organism and can be found in both fresh and saltwater around the world (Fig. 2.3). Conversely, *M. kansasii* has not been found in natural water sources nor soil but has been found in tap water in cities that have prevalent cases [1].

Atypical *Mycobacterium* can present in different manners, specifically MAI can present in three clinically different ways. First, it can resemble typical *Mycobacterium tuberculosis* (MTB). This presentation of MAI is usually seen in middle aged to elderly white men with symptoms that can include cough, weight loss, and upper lobe infiltrates and cavities. The second clinical presentation is nodular bronchiectasis, which has no gender specific association and is typically seen in patients that have no history of smoking, but may have a history of MTB, cystic fibrosis, or other causes of chronically damaged lungs. Finally, the third clinical presentation of MAI is Lady Windermere Syndrome, which is quite interesting and historical. Lady Windermere Syndrome is most commonly seen in immunocompetent, elderly, low body mass index (BMI), nonsmoking women with no preexisting lung disease. These women demonstrate interstitial patterns on a chest X-ray. Lady Windermere syndrome was coined after the character in *Lady Windermere's Fan, A Play About a Good Woman* by Oscar Wilde (Fig. 2.4) due to women in the Victorian-era believing coughing and spitting were ill-mannered acts; by not clearing their lungs they developed a respiratory disease. Although not a part of this discussion, MAI also occurs in the immunocompromised host, such as patients with HIV/AIDS. NTM is commonly thought to be most prevalent in HIV/AIDS patients due to their immunocompromised state; however, the majority of pulmonary NTM seen today is in immunocompetent patients.

To a public health professional, NTM and MTB present different problems. NTM is widespread and can be a severe disease, but unlike MTB, NTM is not transmitted from person to person; instead, it is transmitted through environmental sources via inhalation of the *Mycobacterium* through the respiratory tract or ingestion through the gastrointestinal tract. After entering the

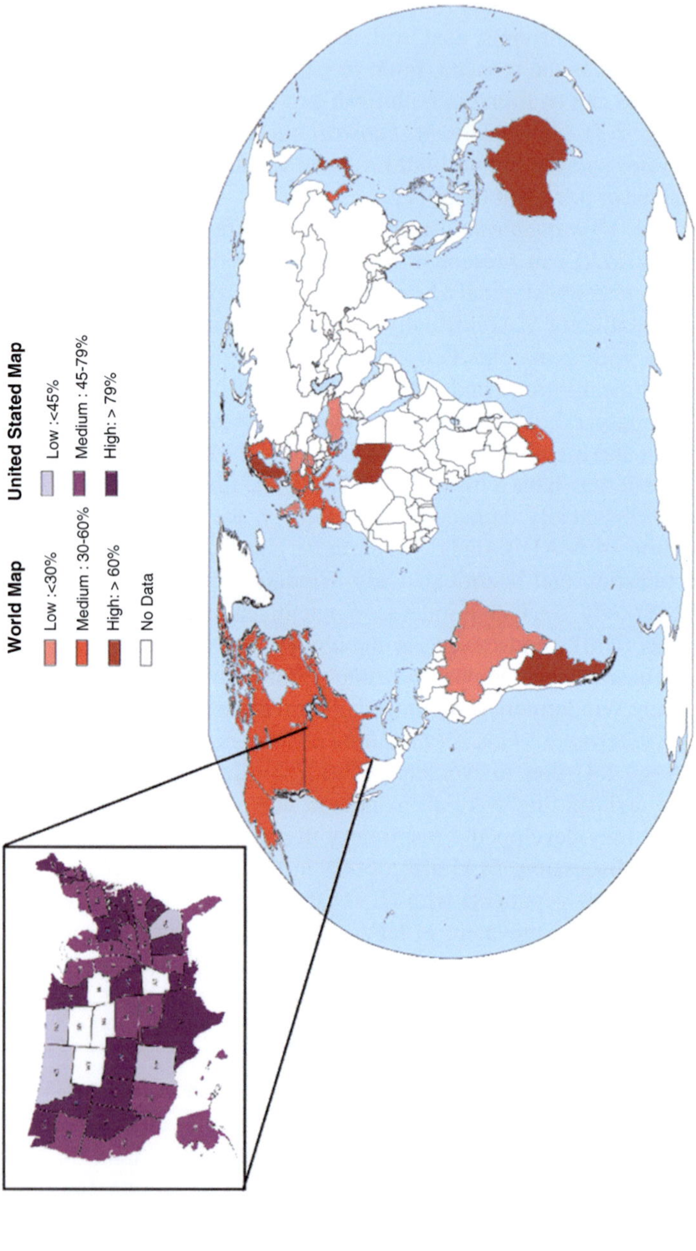

Fig. 2.3 Geographic distribution of *Mycobacterium avium* Complex. World distribution data depicting percentage of MAC found from 2008 pulmonary samples [2]. United States distribution data depicting the percentage of people testing positive for MAC relative to the total number of people testing positive for any NTM infection [3]. Blank World [4] and US [5] map retrieved from Wikimedia Commons

Fig. 2.4 Ad for the 1919 film release of the Lady Windermere's Fan in the USA [6]

body, MAI proceeds to infect pulmonary macrophages and spreads via the lymphatic system. In terms of non-specific symptoms, pulmonary MAI infection manifests in immunocompetent hosts, typically with weight loss, cough, sputum production, fever, lethargy, and night sweats. Other non-specific symptoms may include malaise, dyspnea, chest discomfort, and hemoptysis. Many patients will only present with a chronic cough with purulent sputum production. These symptoms may last weeks to months.

When a patient comes in with symptoms that could be consistent with mycobacterial disease, there are several steps before diagnosis of NTM can be made. In terms of radiographic diagnosis, a chest X-ray may demonstrate cavitary disease in the lung. A CT scan is more sensitive than a routine chest X-ray and is more likely to show other changes such as nodular bronchiectasis. High-resolution computerized tomography (HRCT) scan can show multifocal bronchiectasis and multiple small nodules. In radiographic images, NTM and MTB can both present themselves as apical diseases of the lungs, although NTM is frequency seen in the right middle lobe and lingular segments, which are also the most common areas to find bronchiectasis. This may be related to the angle of the right mainstem bronchus which facilitates aspiration.

In terms of bacteriological diagnosis, an acid fast bacilli (AFB) smear such as the Ziehl-Neelsen acid fast stain would be conducted to determine presence of *Mycobacterium* (Fig. 2.5). Blood cultures are only useful in patients with HIV in determining the presence of *Mycobacterium*; however, analysis must be directly requested as there is a distinct media, and enough time needs to be allowed for growth to occur. Only after 6 weeks of allowing for growth in sputum or blood to occur can a culture be deemed negative with no evidence of *Mycobacterium* present [1]. If mycobacteria are detected, further diagnostic tests are needed to determine which *Mycobacterium* (NTM or MTB) is present. Diagnosis is reliant on analysis of sputum, lung tissue, or other tissues.

While AFB smear is useful to aid in direction of diagnosis, a diagnosis of MAI can only be made with isolation of a sputum culture and nucleic acid amplification (PCR) or with the use of nucleic acid probes. In order to diagnose a patient with MAI, two or more sputum cultures must be positive, or at least one culture-positive bronchial washing or lavage or tissue sample. A patient could also be diagnosed via a biopsy with histopathologic features

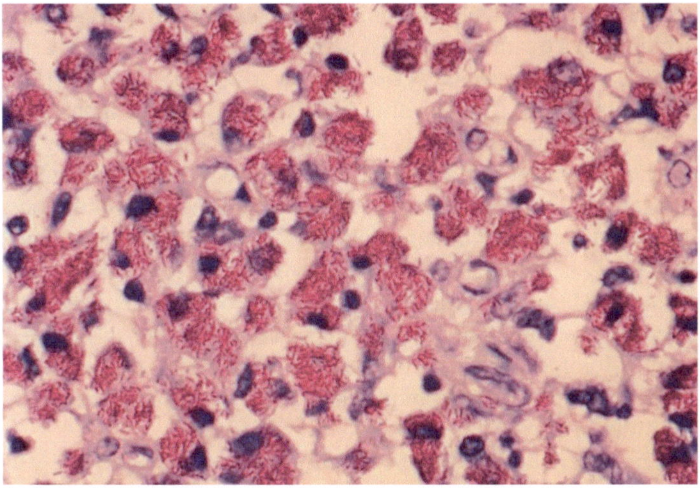

Fig. 2.5 Ziehl-Neelsen Acid Fast Stain Positive for *Mycobacterium* [7]

consistent with mycobacterial infections such as granulomatous inflammation or a positive AFB stain and a positive culture result. Most laboratories testing for MAI currently use highly specific nucleic acid probes which can identify the presence of MAI and yield results within a day of noticeable growth [1].

Treatment requires a holistic review of the patient due to the difficulty, cost, and possible side effects of treatment. For example, in a patient with a positive AFB smear and no symptoms, treatment would not be decided until results of the culture came back. On the contrary, a patient with a positive AFB smear and symptoms such as fever or weight loss may be treated for MTB, due to the infectious nature of MTB, while they wait for the results of the culture. Once the culture results come back, the treatment could stay the course if it revealed MTB or be modified if it revealed MAI. MTB and MAI treatments are similar; therefore, there is no real harm in starting a patient who has MAI with anti-MTB treatments. The treatment benefits need to outweigh the possible side effects.

Spontaneous culture conversion has occurred in about 40–50% of cases, demonstrating not all cases require treatment. To prevent avoidable treatment, physicians should assess several factors before deciding on a treatment path. These include: (1) the patient is at risk for disease progression; (2) has cavitary lesions; (3) low BMI, (4) poor nutrition; or (5) a positive AFB smear. Continual monitoring of the disease progression is important, as a patient originally not needing treatment could later develop the need for medication [8].

Many NTM strains are resistant to conventional antibiotics; thus, the treatment regimen consists of multiple antibiotics, most commonly a combination of three or more antibiotics. This treatment may last as long as 2 years. As stated previously, the side effects of multiple antibiotics could cause difficulty in management. Monitoring of patients receiving treatment is imperative. MAI can be difficult and require three to five medications to successfully treat [9].

The American Thoracic Society/Infectious Diseases Society of America (ATS/IDSA) recommends treatment using a three-drug combination regimen. This three-drug combination includes a macrolide (either clarithromycin or azithromycin), rifampin, and ethambutol. If this combination is not effective or the patient is found to have progressive disease despite treatment, the regimen can be bolstered with an injectable drug of streptomycin or amikacin. Depending on the patient's case, the treatment regimen can consist of taking the prescribed medications three times per week or daily. The recommended duration of treatment is to continue therapy until 12 months of continuous sputum cultures are negative [8]. This means frequent sputum cultures, starting about 6–9 months after treatment initiation and about every 3–6 months, are done during treatment. In rare cases, surgical management via an adjuvant lung resection surgery may be used for those with localized cavitary disease. This is usually for those with drug intolerance or failure and requires an exhaustive risk-benefit assessment and a very experienced surgeon [8].

Diagnosis

On all three occasions Agatha had extensive workup for infection and malignancy. The workup for malignancies all came back negative. Chest X-rays and CT scans of the chest revealed cavitation and nodules as well as three sputum cultures tested positive for MAI with moderate acid fast bacilli (AFB) on a smear. She was diagnosed for the second time with MAI. On all three occasions NTM was discovered: the first time was *M. kansasii*, and the second and third time was MAI. Diagnosis was made by sputum evaluation and bronchoscopy which demonstrated the presence of *Mycobacterium*.

Treatment

In Agatha's case, she was originally treated for *M. kansasii* for 18 months with INH, Rifampin, Ethambutol until she presented with radiographic resolution and negative cultures. Following her treatment, she demonstrated no pulmonary symptoms. After about 10 years of being symptom free, she was diagnosed with MAI for the first time and treated for 22 months with Azithromycin, Rifampin, and Ethambutol. Azithromycin, instead of Biaxin, was used in Agatha's case due to her intolerance of Biaxin. Following treatment, there was complete resolution of symptoms, but based on radiographic images, residual cavitation in the left upper lobe was present. When she was diagnosed for the second time with MAI 3 years later, she was treated once again with the three-drug combination regimen of Azithromycin, Rifampin, and Ethambutol. One could argue her second MAI infection was due to failure to resolve her first MAI infection, and this subsequently resulted in re-infection. It could also be argued that because she had it 3 years prior, she required a more aggressive treatment the second time.

Conclusion

Agatha is a 72-year-old woman with 3 different NTM diagnoses, starting with *M. kansasii*, MAI for the first time, and then MAI again for the second instance. MAI is the most common NTM, followed by *M. kansasii*. Diagnosis is a multi-step process. Treatment is long and rigorous, consisting of several drugs and continuing until patients have negative sputum results for a year.

Fun Facts

- Clinical observations in non-smokers versus smokers slightly differ; nodular bronchiectasis is more common in non-smokers and cavitary disease is more common in smokers.
- Studies have shown states with a higher water vapor content have a higher burden of NTM infection [9].
- Pectus excavatum (sunken breastbone), scoliosis, mitral valve prolapses [1], and hypersensitivity pneumonitis (hot tub lung) are clinical observations correlated with diagnosis of MAI.

Multiple Choice Questions

1. Nontuberculous mycobacteria (NTM) infections are considered contagious.
   (a) True
   (b) False

2. Who is most likely to be diagnosed with Lady Windermere syndrome?
   (a) Young, Caucasian, female
   (b) Elderly, Asian, male
   (c) Elderly, Caucasian, female
   (d) Elderly, African American, female

3. A positive AFB smear automatically indicates a patient has MAI.
   (a) True
   (b) False

4. If a three-drug treatment regimen has not proven to be effective, what should a physician do?
   (a) Double the dose of current medication
   (b) Perform an adjuvant lung resection surgery
   (c) Strengthen the regimen with streptomycin or amikacin
   (d) Stop treatment

Answers

1. (b)
2. (c)
3. (b)
4. (c)

References

1. Griffith DE. Diagnosis of nontuberculous mycobacterial infections of the lungs. 2020. https://www.uptodate.com/.
2. Hoefsloot, et al. The geographic diversity of nontuberculous mycobacteria isolated from pulmonary samples: a NTM-NET collaborative study. Eur Respir J. 2013;42:1604–13. https://doi.org/10.1183/09031936.00149212.
3. Mirsaeidi M, Vu A, Leitman P, Sharifi A, Wisliceny S, Leitman A, et al. A patient-based analysis of the geographic distribution of Mycobacterium avium complex, Mycobacterium abscessus, and Mycobacterium kansasii infections in the United States. Chest. 2017;151(4):947–50. https://doi.org/10.1016/j.chest.2017.02.013.
4. Dlouhý P. File: world map blank—with blue sea.svg [Digital image]. 2006. https://commons.wikimedia.org/wiki/File:World_Map_Blank_-_with_blue_sea.svg. Accessed 23 Sep 2020.
5. Fredericks C. File: blank US map borders labels.svg [Digital image]. 2007. https://commons.m.wikimedia.org/wiki/File:Blank_US_map_borders_labels.svg. Accessed 23 Sep 2020.
6. Triangle Distributing Corp. File: Lady Windermere's Fan (1916)—Ad 1.jpg [Ad for the British film Lady Windermere's Fan (1916), for its 1919 release in the United States, on page 1422 of the June 7, 1919 Moving Picture World]. 1919. https://commons.wikimedia.org/wiki/File:Lady_Windermere%27s_Fan_(1916)_-_Ad_1.jpg. Accessed 24 Apr 2020.
7. Ewing EP Jr., Centers for Disease Control and Prevention. File: Mycobacterium avium-intracellulare.jpg [Mycobacterium avium-intracellulare infection of lymph node in patients with AIDS. Ziehl-Neelsen stain. Histopathology of lymph nodes shows tremendous numbers of acid-fast bacilli within plump histiocytes]. 2007. https://commons.wikimedia.org/wiki/File:Mycobacterium_avium-intracellulare.jpg. Accessed 24 Apr 2020.
8. Kwon Y-S, Koh W-J, Daley CL. Treatment of Mycobacterium avium complex pulmonary disease. Tuberc Respir Dis. 2019;82(1):15. https://doi.org/10.4046/trd.2018.0060.
9. Kasperbauer SH. Nontuberculous Mycobacteria (NTM) overview. 2017. https://www.nationaljewish.org/conditions/ntm-nontuberculous-mycobacteria/ntm-nontuberculous-mycobacteria-overview.

"I'm Tired and in Pain!" When Small Things Make a Big Difference

3

Zeyu Tang

Z. Tang (✉)
Yale University School of Medicine, New Haven, CT, USA
e-mail: jtang1@ualberta.ca; zeyu.tang@yale.edu

A. M. Szema (ed.), *Unusual Diseases with Common Symptoms*,
https://doi.org/10.1007/978-3-031-40280-7_3

Milton Glen is a 52-year-old woman who was referred to Dr. Szema for "pain over my entire body" and constantly feeling more tired than usual.

A few months ago, she visited a neurologist for some non-specific symptoms that she had never experienced before. Her husband was actually the first one to notice her symptoms—she started kicking him in her sleep.

Over the next few weeks, her sleep quality progressively declined, which drained her energy, mood, and concentration. Other symptoms began to develop as well. She had deep muscle aches all over her body. Standing up, walking, and performing the most menial tasks made her dizzy—to the point that she had to take breaks in between gardening for the backyard to wait for the world to stop spinning. Occasionally, her heart would often start pounding and she would sweat and shake, as if someone had just injected her with adrenaline. Worst of all, *she just felt sick all the time.*

Dr. Szema contemplated. One of the most common problems for a woman at this age is menopause. The signs and symptoms of perimenopause are numerous and highly variable from person to person. Perimenopause could fit. But Milton Glen mentioned to Dr. Szema on routine questioning that she was already in menopause, and it had been 2 years since her last symptoms. Even more than that, this did not feel the same. No previous treatment with hormone replacement therapy, either, which, if taken, could have caused her symptoms was identified. Perhaps it is not menopause, then.

Dr. Szema looked back to her original referral note. Her neurologist had written down Shy-Drager syndrome (Fig. 3.1) with two question marks—an uncertain diagnosis, but a troubling one if the neurologist's initial assessment was right.

A diagnosis of Shy-Drager corresponded with nearly all of her complaints, but pain is not a typical feature of Shy-Drager. There was still nothing that explained her pain. The neurologist took a thorough interview and examined Milton Glen head to toe—nothing was out of the ordinary. After the neurologist had excluded any organic causes of her constant body aches, he gave her a clinical diagnosis of exclusion: fibromyalgia. Fibromyalgia is a chronic disease characterized by widespread pain in the muscles, ligaments, and tendons. What makes fibromyalgia special is that there is nothing from the patient's history, physical exam, or laboratory tests that could be identified as a cause for the pain. Tissue biopsies of patients with fibromyalgia will show no abnormalities and no evidence of inflammation. Given only when doctors have ruled out any other possible causes (Fig. 3.1).

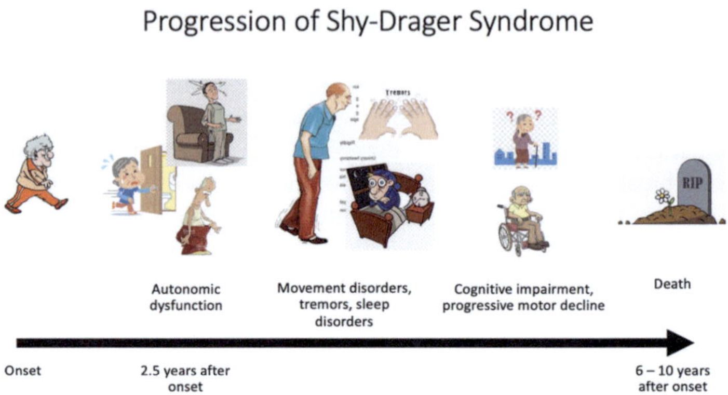

Fig. 3.1 Shy-Drager syndrome. Shy-Drager syndrome is a nervous system disease that first appears with dizziness, mood changes, problems sleeping, and a dysfunctional autonomic nervous system [1]. It is classified as a neurodegenerative disease, a broad category of diseases which means that something is causing neurons to die. Moreover, Shy-Drager is classified as a "Parkinson-plus" disease because it has the features of Parkinson's disease with additional features. As with all other neurodegenerative diseases, there is no cure, and the disease becomes progressively worse and ultimately leads to death. The majority of patients die within 10 years after the symptoms appear

Despite how common, debilitating, and extensively researched fibromyalgia is, we still do not have a clear understanding of what it is caused by. Could Shy-Drager be related to Milton Glen's fibromyalgia? Both are disorders of the brain, both are poorly understood, both appeared in Milton Glen at around the same time. It would not be far-fetched to think they were somehow connected. However, there have only been a handful of reports of the two diseases together in the medical literature. Physicians are trained to be vigilantly wary of their own biases—human psychology tends to favor explanations that tie everything together neatly rather than accepting coincidences. But in reality, with the lack of evidence, Milton Glen's two conditions could be interpreted as coincidence rather than a single unified disease.

The neurologist prescribed gabapentin and duloxetine, two medications that alter nerve function and are the classic first-line therapy for fibromyalgia. Milton Glen took the medicine as directed, but by the end of the 4 grueling weeks, nothing had changed. She returned to the neurologist, who referred her to Dr. Szema for an in-depth investigation into her pain.

Dr. Szema is a specialist in adult and pediatric allergy/immunology and pulmonary disease. But in more than 20 years of practice, he has also gained a reputation in the medical community for working with patients with serious but ambiguous problems. He is a doctor, but sometimes he works more like a medical detective. Dr. Szema asked Milton Glen to tell her story from the beginning. Her aching pains first started after being hospitalized for severe pneumonia. At that time, her pain was mild and tolerable, so she tolerated it with the expectation that it would resolve on its own. Instead, the pain gradually worsened, and she started to notice increased fatigue throughout the day.

It is important that doctors gain as much information from the patients as possible, even when it seems like there is nothing left for the patient to mention. "Anything else that you can remember?" Dr. Szema asked. She finally noted that her sinuses have also been more congested recently: "it seems like I'm getting a new cold every other day now." Aside from Shy-Drager syndrome, her medical history included a benign nodule on her adrenal gland and chronically low blood pressure. She added she was allergic to peanuts and thought she might have seasonal allergies.

This is where most clinicians would wrap up their conversation. In the consideration of time and efficiency, most clinicians rely on what they consider as the high yield questions—questions that have the highest chance of obtaining diagnostically relevant information while de-emphasizing the questions that have a low chance of being important. High yield questions include, "Has anyone else in your family ever experienced this before?" and "any fevers or chills recently." Most doctors do not ask patients if they use air conditioning.

Dr. Szema has a different approach, as he is often the last line for patients seeking answers. He focuses on topics that others have overlooked and leaves no stone unturned. As a result, his diagnoses involve factors that would seem harmless to most. Dr. Szema carried on with his questioning.

Dr. Szema asked about her sleep with fervent attention to the details. Milton Glen slept in the same bed with her husband and four dogs. Her blankets were made from cotton, and her pillows had feather stuffing. He pressed for details—what were the materials used for her bed? Feather pillows and cotton blankets. How often do you clean your drapes? She looked at him, stunned, wondering whether he was being serious. "Never," she replied.

What does the house look like? When was it built? Milton Glen's house was relatively old—built in the 1950s—and Milton Glen could not remember the last time the air filter was changed. Her house has air conditioning, which was continuously running in the hot summer months.

A story began to form. Milton Glen suffered from seasonal allergies even when she was healthy. Dogs are notorious producers of allergens. Cotton and feather objects have the unfortunate tendency to trap allergens. Drapes are a source of allergens that few people think about. Older houses carry more dust and mold, and a dirty air filter would disperse allergens into the air rather than clean it. An air conditioner, with a failed home air filter, can aerosolize and recirculate allergens back into the house. She was constantly exposed to allergens at home, and there had not been any efforts to address the poor air quality. Milton Glen mentioned frequent colds, which could be caused by inflammation of the nasal passages from chronic allergen exposure. Allergens can cause reactive airway disease, where the constriction of the airways reduces air flow into the lungs and negatively impacts the ability for the body to receive oxygen.

Dr. Szema moved on to the physical exam and started by checking her vital signs. Milton Glen had an unusually low blood pressure and a high heart rate, which suggested to Dr. Szema that there was something wrong with her circulatory system. He followed his hunch and tested her with an orthostatics exam (Fig. 3.2).

Milton Glen failed the orthostatic exam, which indicated to Dr. Szema that her sympathetic nervous system was not responding normally. This was likely due to her Shy-Drager syndrome, which is associated with defects in the autonomic nervous system. Having an abnormally low blood pressure can mean that the brain is being deprived of blood flow than it needs to perform optimally.

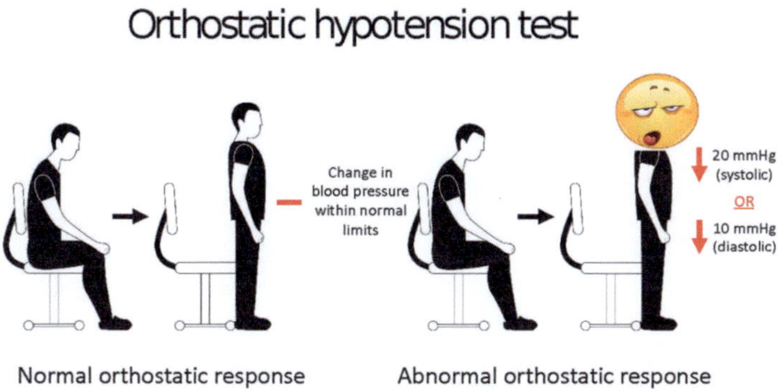

Fig. 3.2 Orthostatic hypotension test. An orthostatics exam is a non-invasive test to see how the body responds to being in different positions. Normally when a person stands up, blood from the legs and feet have a harder time fighting gravity to reach the heart. If nothing was done, blood would quickly begin to pool in your lower body. Blood pressure would be in the upper body, and there would be a loss of blood flow to vital organs such as the heart and brain. Our bodies have adapted to this challenge to preserve blood flow when standing. As soon as a person stands, their brains sense the decreased blood pressure and reflexively signals to constrict the blood vessels in the lower extremities. Blood that would have pooled in the legs forced out of the lower limbs and back to the heart. But in some people, this reflex is weakened, which can cause lightheadedness, dizziness, blurred vision, fatigue, and problems concentrating. People can be at risk of passing out just from standing up

Dr. Szema ordered a comprehensive battery of preliminary tests before their first meeting, which included a test for Lyme disease and an erythrocyte sedimentation rate. The Lyme disease test indicated that there was no active infection, but showed there was exposure to Lyme disease in the past. Milton Glen denied any Lyme symptoms but did endorse going on the occasional hike. "Any contact with wild animals?" Milton Glen thought for a minute. "I did work with wild animals in my previous job, I quit about 2 months before these symptoms started." Even when a Lyme infection has run its course, people exposed to Lyme disease are at risk of "Post-Lyme disease syndrome," which is a constellation of non-specific symptoms all around the body—including fatigue. Dr. Szema looked at the erythrocyte sedimentation rate, a non-specific indicator of inflammation in the body (Fig. 3.3).

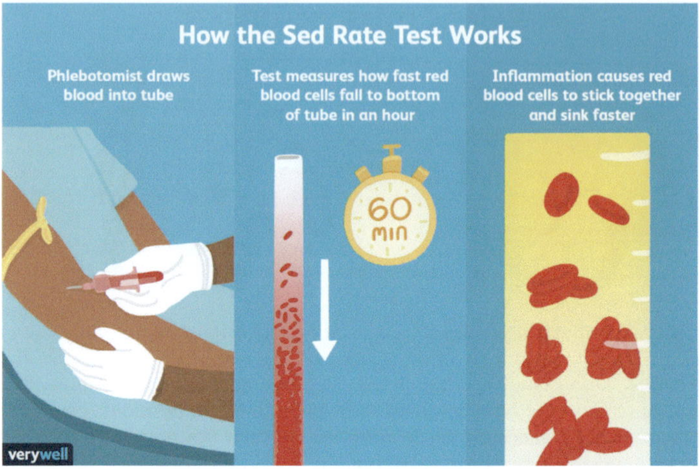

Fig. 3.3 Sed rate test

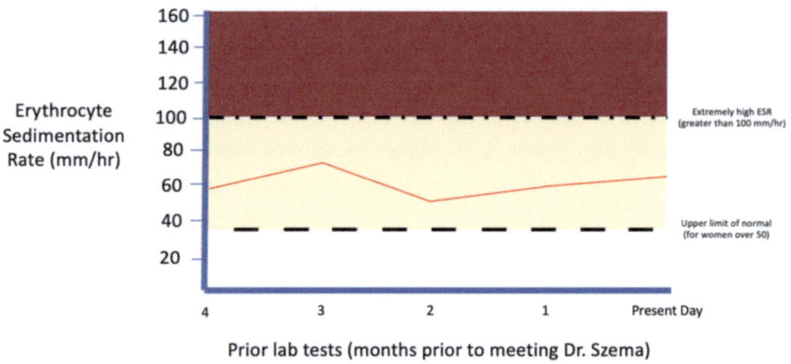

Fig. 3.4 Milton Glen's previous erythrocyte sedimentation rates

It was suspiciously elevated. Dr. Szema looked back at Milton Glen's previous tests and found that the ESR had been consistently high for the past few months, suggesting a chronic inflammatory process. Something—whether it was her allergies, or post-Lyme disease syndrome, or some other unknown condition—was causing her body to be inflamed (Fig. 3.4).

Fibromyalgia, chronic exposure to allergens, low blood pressure, post-Lyme disease syndrome and inflammation can all cause fatigue. Fatigue is a well-known trigger for the pain symptoms

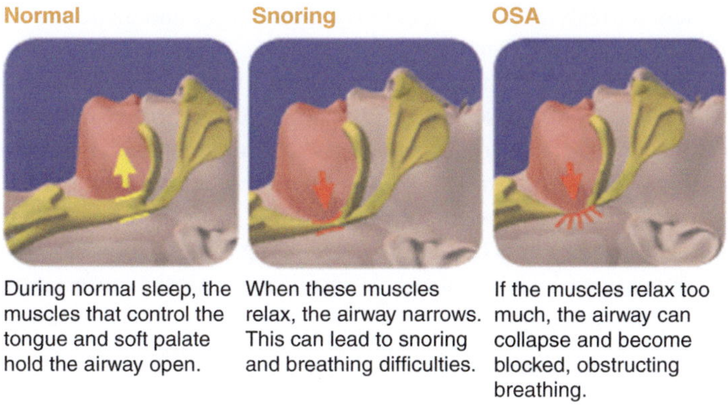

Fig. 3.5 Sleep apnea

of fibromyalgia. Dr. Szema was still skeptical—none of these explains why she is kicking her husband in her sleep. Dr. Szema sent Milton Glen for a sleep study. After being monitored for a night in the sleep lab, Milton Glen was found to have mild sleep apnea (Fig. 3.5), with 28 separate instances of hypopnea, abnormally shallow or slow breathing. They also confirmed that she was kicking in her sleep: 15 times per hour and had abnormally increased movements throughout the night.

Diagnosis

Chronic fatigue caused by fibromyalgia, mild obstructive sleep apnea, periodic limb movement, upper airway resistance syndrome, neurogenic orthostatic hypotension from Shy-Drager syndrome, allergic rhinitis, and possible exposure to Lyme disease.

Salient Features

This case highlights the multimodal approach to investigating and treating a patient with multiple causes for fatigue. There are countless reasons for why someone might be tired all the time, and many of them can be managed through simple changes to the patient's lifestyle. Here we identify each trigger and elaborate on the recommended therapy.

Shy-Drager syndrome or multi-system atrophy (MSA) is a neurodegenerative disease that results in the destruction of the parts of the brain that control for movement, coordination, and the autonomic nervous system [1]. At this time there is no cure, and the primary treatment is symptom relief through physical rehabilitation and medications that improve signaling of the neurotransmitter dopamine (dope-ah-meen). A significant clinical manifestation of Shy-Drager syndrome is postural hypotension, where the autonomic nervous system is unable to maintain blood pressure in upright positions such as standing or exercising. A chronically low blood pressure can lead to the feeling of fatigue and cognitive slowing simply because there is less blood perfusing the brain. Pharmacologic treatments target two different components of blood pressure: retaining more fluid within blood vessels and improving autonomic nervous system responsiveness, which are enacted by steroids fludrocortisone and alpha-adrenergic agonists such as midodrine, respectively. There are also important non-pharmacological ways to increase blood pressure, such as taking in more salt in diet and wearing compression stockings that force more blood to the head.

In this case, Milton Glen's sleep disturbances can also be connected to her Shy-Drager syndrome. Normally when we sleep, we suppress the signaling between our brains and muscles. This allows us to imagine ourselves moving in our dreams without moving in the real world. In some diseases there is a lack of signal suppression, which causes patients to act out their dreams. This can manifest as talking, sleep, sleepwalking, or performing other tasks during sleep. In Milton Glen's case, the condition manifested as flailing her limbs in her sleep which caused her to occasionally kick her husband (Fig. 3.6 and Table 3.1).

Milton Glen's tests also suggested that she was exposed to Lyme disease at some point in her life. Post-Lyme disease syndrome presents with any number of subjective and non-specific symptoms such as fatigue, widespread musculoskeletal pain, and cognitive decline [5]. Lyme disease is most often associated with patients who traveled or lived in the Northeastern United States,

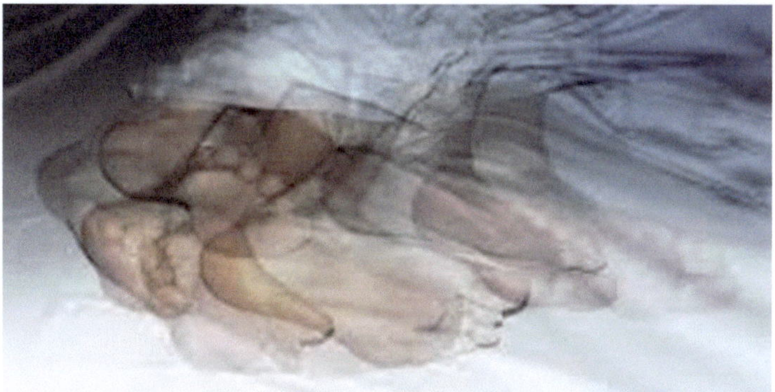

Fig. 3.6 Periodic limb movements during sleep

which is the case for Milton Glen. Despite the classic localization to the New England area, Lyme disease has been found all over the world: in the Northern hemisphere, in central Europe, Asia, and North America. There have also been case reports from Northern Africa and South America. The disease is severely underdiagnosed, and many are unaware of their previous exposure. About 30,000 cases of Lyme disease in the USA are reported to the Centers for Disease Control and Prevention (CDC) each year, but the CDC notes the actual number of cases could be upward of 300,000 new cases per year [6].

Obstructive sleep apnea can cause tiredness and fogginess throughout the day even if the patient has a normal oxygen saturation when awake. Studies have found that women with fibromyalgia may also have a high likelihood of breathing disorders in their sleep [7]. One hypothesis states that problems with breathing during sleep are involved in the development and aggravation of fibromyalgia [7].

Obstructive sleep apnea is very common in the general population with an estimated one in five adults affected and is associated with aging and obesity [8]. As such, the incidence has been steadily increasing over the last few decades due to the growing obesity epidemic. At a normal body mass index of 23, Milton Glen does not have the classic risk factor associated with obstructive sleep apnea. Even so, a surprising one in four people with sleep

Table 3.1 MSA vs. Parkinson's vs. Lewy body dementia vs. MS

Disease	Epidemiology	Pathophysiology	Clinical presentation	Treatment
Parkinson's disease [2]	Common	Dopaminergic neuronal degeneration in substantia nigra in brain	"Parkinsonism": Bradykinesia, resting tremor, rigidity	Medications to increase dopamine
			Shuffling gait	Deep brain stimulation
			Sleep disorders	Physical therapy
			Dementia (late stage)	Evaluation for depression
				Evaluation for sleep disorders
Multiple system atrophy	Rare (prevalence of 2=5 people per 100,000)	Dopaminergic neuronal degeneration in substantia nigra in brain	Symptoms of Parkinson's disease ("Parkinsonism")	Medication to increase blood pressure
			Autonomic dysregulation	Physical therapy
			Sleep disorders	Evaluation for depression
			Dementia (late stage)	Evaluation for sleep disorders
Lewy body dementia [3]	Common (second most common cause of neurodegenerative dementia)	Build-up of Lewy bodies (alpha-synuclein-positive particles) in glia cells in brain	Visual hallucinations	Behavioral therapy
		Cortical atrophy of brain	Episodes of paranoia	Physical therapy
			Dementia	Medications for parkinsonian symptoms
			Sleep disorders	Medications for dementia symptoms

(continued)

Table 3.1 (continued)

Disease	Epidemiology	Pathophysiology	Clinical presentation	Treatment
Multiple sclerosis [4]	Common (prevalence of 100=150 people per 10,000 in the USA)	Autoimmune disease leading to demyelination and inflammation in the central nervous system	Fatigue	Steroids during acute exacerbations ("flares")
			Depression	Various immune-modifying medications to prevent flares
			Generalized pain	
			Vision problems such as temporary vision loss and double-vision	
			Motor weakness and coordination and balance problems	
			Electric shock-like sensations running down the spine when flexing the neck (Lhermitte sign)	
			Sensitivity to heat (Uhthoff phenomenon)	
			All of the above are relapsing symptoms	

apnea is non-obese [9]. Non-obese OSA patients tend to be more difficult to treat than obese patients, but the same treatment recommendations are given for both groups: use of breathing assistance devices such as continuous positive air pressure (CPAP) machine and weight loss. In fact, CPAPs are so safe and effective for treating obstructive sleep apnea that experts have pushed for all patients with severe obstructive sleep apnea to be treated with CPAP [10]. In addition to treatments, patients with obstructive sleep apnea are cautioned to avoid alcohol, sedative sleep medications, and opioids, as these depress central respiratory signaling and can exacerbate problems with breathing during sleep.

Unfortunately, CPAP machines are expensive, cumbersome to use, and disruptive for partners sleeping next to patients. For these reasons, CPAPs have been found to have adherence rates of 13–71%, even though they are effective treatments for obstructive sleep apnea [11]. Recent advances in obstructive sleep apnea management have resulted in effective alternatives to CPAP devices. For example, a landmark 2014 study published in the New England Journal showed that obstructive sleep apnea could be effectively treated with surgically implanted pacemakers [12]. These pacemakers stimulate the hypoglossal nerves, nerves that can move the tongue out of the airway, when the device senses a breath being taken during sleep. They work best with non-obese obstructive sleep apnea patients, since the fundamental defects in this group are more likely to be related to tongue obstruction rather than airway collapse (Table 3.2).

Conclusion

After going over the sleep study results, Dr. Szema prescribed medicines and lifestyle changes for Milton Glen. Initially, her treatments seemed to be working. One medication he prescribed, Northera, stabilized her blood pressure. He ordered a continuous positive airway pressure (CPAP) machine for her which improved

Table 3.2 Plan of action

Cause of fatigue	Management
Fibromyalgia	1. Pain relief: neurontin, cymbalta
	2. Address causes of fatigue, which trigger fibromyalgia
Obstructive sleep apnea	Weight loss, CPAP machine, dental devices, hypoglossal pacemaker
Periodic limb movements during sleep	Drugs that suppress muscle contractions
Upper airway resistance syndrome and allergic rhinitis	Weight loss
	Identify common environmental allergens: dust, mold, pollen, pet dander
	Address environmental allergens: dust mite proof mattress covers, pillow case covers, cleaning home surfaces that are often overlooked, HEPA air filter, home furnace inspection and filter replacement, not letting pets sleep on bed
	Avoid alcohol
Neurogenic orthostatic hypotension from Shy-Drager syndrome	Droxidopa (drug)
	TED compression stockings
Post-Lyme disease syndrome	None, usually disappears within 6 months after Lyme infection is cured

Dr. Szema's treatment plan for Milton Glen

her sleep quality and greatly reduced fatigue. In their last appointment, she was cheerful, pain free, and back to her normal energy levels.

But when Dr. Szema reached out for an update years later, Milton Glen stated that the chronic fatigue returned. She found a new job, one that provided different insurance than the one she had before. Her new insurance company refused to cover the use of CPAP for her apnea and discontinued her Northera. At the present she is back on Northera and remains mildly fatigued throughout the day.

Fun Facts

① Multiple system atrophy, previously known as Shy-Drager Syndrome is a progressive, systemic failure of multiple organs, beginning with the autonomic nervous system (excessive sweating, postural hypotension), sexual dysfunction, bladder and bowel incontinence, and dementia.

② Living in an endemic region such as Long Island, NY, and CT during the temperate seasons with associated constellation of symptoms—lassitude, malaise supports the concept of testing for Lyme disease.

③ Occam's Razor may not apply in 2020 with all the causes of fatigue. There can be many different causes superimposed onto each patient, and each of them may require a different treatment modality.

Multiple Choice Questions

1. What is the main difference between Parkinson's disease and multiple system atrophy/Shy-Drager syndrome?
   (a) Presence of significant autonomic dysfunction in multiple system atrophy
   (b) Tremors are absent in multiple system atrophy
   (c) Parkinson's disease is a central nervous system disorder, while multiple system atrophy is a peripheral nervous system disorder
   (d) Multiple system atrophy is curable, unlike Parkinson's disease

2. Which of these *is not* a known cause of chronic fatigue?
   (a) Type 2 diabetes
   (b) Chronic cocaine use
   (c) Daily afternoon coffee
   (d) Use of opioids, sedatives, or sleeping pills
   (e) Working night shifts or rotating shifts
   (f) All of the above can cause chronic fatigue

3. Which region of the world is Lyme disease most endemic to?
   (a) Northeast United States
   (b) Central China

(c) Eastern Europe
(d) South America

Answers

1. (a)
2. (f)
3. (a)

References

1. Wenning GK, Geser F, Krismer F, et al. The natural history of multiple system atrophy: a prospective European cohort study. Lancet Neurol. 2013;12:264–74.
2. Clinical manifestations of Parkinson disease. 2020. Accessed 8 Aug 2020.
3. Clinical features and diagnosis of dementia with Lewy bodies. 2017. Accessed 8 Aug 2020.
4. Oh J, Vidal-Jordana A, Montalban X. Multiple sclerosis: clinical aspects. Curr Opin Neurol. 2018;31:752–9.
5. Novak P, Felsenstein D, Mao C, Octavien NR, Zubcevik N. Association of small fiber neuropathy and post treatment Lyme disease syndrome. PLoS One. 2019;14:e0212222.
6. Data and surveillance. 2019. Accessed 8 Aug 2020.
7. Gold AR, Dipalo F, Gold MS, Broderick J. Inspiratory airflow dynamics during sleep in women with fibromyalgia. Sleep. 2004;27:459–66.
8. Franklin KA, Lindberg E. Obstructive sleep apnea is a common disorder in the population—a review on the epidemiology of sleep apnea. J Thorac Dis. 2015;7:1311–22.
9. Gray EL, McKenzie DK, Eckert DJ. Obstructive sleep apnea without obesity is common and difficult to treat: evidence for a distinct pathophysiological phenotype. J Clin Sleep Med. 2017;13:81–8.
10. Pack AI, Gislason T. Obstructive sleep apnea and cardiovascular disease: a perspective and future directions. Prog Cardiovasc Dis. 2009;51:434–51.
11. Bakker JP, Weaver TE, Parthasarathy S, Aloia MS. Adherence to CPAP: what should we be aiming for, and how can we get there? Chest. 2019;155:1272–87.
12. Strollo PJ Jr, Soose RJ, Maurer JT, et al. Upper-airway stimulation for obstructive sleep apnea. N Engl J Med. 2014;370:139–49.

Eosinophilic Colitis: Unforeseen Consequence of the Vietnam War

4

JacqueLyn Kinney

J. Kinney (✉)
Hackensack Meridian School of Medicine, Nutley, NJ, USA

Plastic and Reconstructive Surgery Division, Department of Sugery, Rutgers University, Newark, NJ, USA
e-mail: jacquelyn.kinney@hmhn.org; jrk427@nyu.edu

A. M. Szema (ed.), *Unusual Diseases with Common Symptoms*,
https://doi.org/10.1007/978-3-031-40280-7_4

Introduction

Doc was a troubled gentleman who came to see Dr. Szema after struggling to find a cure to his chronic stomach pain. Previously, he had been treated for what his medical provider thought to be **ulcerative colitis**. His team tried everything to help Doc get back to his normal life—even going as far to perform a colectomy, which involves removing his colon and leaving an ostomy bag (to hold his feces) hanging off his stomach. This is a difficult decision to make for anyone, but the thought that it would help cure him of his pain, made it all worth it … However to their dismay, his pain never resolved. At this point, he had traveled to several providers, bringing his medical records and stating that he had ulcerative colitis. This anchored his diagnosis in his provider's minds to ulcerative colitis. Since they had already been primed to believe this patient had ulcerative colitis, this made it hard to look past and consider different rare diagnoses that could explain why his pain never resolved with typical treatments.

This is the story of how Doc found his pain relief through further evaluation, and how being anchored to a certain diagnosis could do more harm than good. It is for this reason that we must continue to reevaluate each patient that comes into our office or clinic to ensure nothing—no matter how rare—gets missed.

Vignette

Once upon a time, there was a young man named Doc who tossed and turned at night, not knowing why his body was seeming to revolt against him at every turn. He had already exhausted every over-the-counter treatment he could think of, and yet had received no relief. At the end of the day, all he wanted was to return to his daily life as a normal person, doing the things he loves, exploring life—not laying in bed writhing in pain.

He had previously served valiantly in the military defending freedom in Vietnam. He was a hero to his country and made his family proud with his hard working spirit. Troubled times hit though, and Doc was suddenly plagued with always being sick, losing weight, not eating. His condition continued to worsen to a chronic disability and led to his inability to work. This defender of freedom was now unemployed, and without health insurance.

(Drawing of an overweight young man in uniform walking through a village, one hand on stomach, the other in his pocket, looking sadly downward).

His life was not always like this—he used to fit in with those he was deployed with and his days off were filled with joy and laughter. He never slacked about and always made sure his work was complete on time. Slowly but surely, his life started changing. At first it started with a strange abdominal pain after he finished a meal. Although inconvenient, that was not a major problem and he was always able to work through the discomfort. It was not long until that discomfort turned into diarrhea and vomiting. Even though he tried to avoid certain foods and take herbal supplements, his symptoms did not seem to get any better. He started to lose interest in eating all together since he knew the symptoms that would soon follow.

He could no longer attend the weekly poker nights his friends put on and he would often need to cancel plans to get off base due to his increasing fatigue. How could he regain his strength if he could not eat? It seemed like everything that once made him happy was causing him pain and discomfort. He wanted nothing more than to go out and rejoin his friends and family, or even return to work.

Doc was a very strong willed, stubborn man, who tried for months to withstand the pain and frequent diarrhea. He began dropping weight at such a dramatic pace that his trousers began falling off of him. He estimated that he must have dropped 100 pounds in just the last year alone! He was in despair when he was no longer able to keep up with his work duties and needed to return home. This disappointment made him even more determined to take his future in his own hands. He was not about to bow out due to an unknown illness. He knew that there were no longer new herbs or self-care he could try that would fix his problem. That is when he decided to embark on a journey of a lifetime.

He jumped on his computer and scoured the internet for medical providers. He was determined to find someone in this world that would have some idea of what was going on inside of him.

His primary care provider told him to travel to both a medium sized public tertiary care hospital and medical school, as well as a regional veteran affairs hospital to see if experienced, specialized physicians could help. Although he knew that it would be difficult to travel and he did not have much money, he was determined to seek out a physician that would be able to get his old life back. He sat down with each provider and explained the massive toll this has had on his life. The first doctor listened and decided to run some tests that Doc did not completely understand. He was more concerned with finding an answer than quizzing the physician on his thought process.

It was this healer that diagnosed him with **ulcerative colitis**. It was the doctor's thought that his diarrhea, abdominal pain, fatigue, and weight loss were pointing toward this fairly common diagnosis. It is estimated that 1.3% of the US population is suffering from ulcerative colitis or Crohn's disease.[1] It would make sense that this common disease that causes *inflammation of the large intestine with open sores* could cause these symptoms. Ulcerative colitis is a type of autoimmune condition, when your immune system does not recognize your intestines, and starts treating it like a foreign invader. Doc returned home with the steroids and aspirin he was prescribed to decrease his body's immune response to his intestines. Doc was aware that the new medications he was on would likely not cure his chronic disease, but would likely cure him of his symptoms, so he felt comfortable returning home in the hopes that he would feel better.

Doc patiently waited at home for the new medications to take effect. He waited … and waited … and waited … However, his symptoms never went away. The physician had given him such high hopes that the medications would be able to give him back his old life. He soon began worrying that something was not working the way it should have. He had lost hope in that doctor and borrowed his neighbor's car to travel to the big city. There, he would meet with a second physician that might uplift his dwindling hope for a cure.

This physician took a look at Doc and decided to run a stool ova and parasite exam (O&P) just to rule out any parasitic infections in his gastrointestinal tract. This ended up being negative, meaning he did not have any parasites living in his bowels. Therefore, the physician took another look at him, and thought once again that he had to have ulcerative colitis. The symptoms Doc presented with and the lack of alternative explanations just seemed to line up with a diagnosis of ulcerative colitis. This phy-

[1] "Data and Statistics"

sician explained to Doc that when patients with ulcerative colitis do not respond to steroids, they are termed "refractory," which means that they are hard to treat. This is when medical teams normally have to take a more aggressive approach on treatments.

Doc was still having the same symptoms and the doctor was growing even more concerned about his nutrition at this point. He had already lost so much weight, and his continued disrupted eating would surely leave him deficient in more than one vitamin. This doctor took a long look at his case, made the heavy recommendation for Doc to get his colon removed—a procedure known as a colostomy.

Since he did not have health insurance, a kind surgeon at a community hospital agreed to do the procedure for him. The colostomy procedure would hopefully be able to remove all of the inflamed parts of his bowels and reroute his fecal matter to a colostomy bag that would hang off his stomach.

With a heavy heart, Doc proceeded forward with the surgery. He knew that this would be life changing—but that is exactly what he needed. He needed a change, to be able to get back to normal life. To return to his friends and family and be able to work again. Ultimately, he needed to be able to eat—regardless of what it cost him.

It was after the colostomy, and after he traveled home, when his troubles restarted. He was convinced that he just needed to wait it out for the procedure to work. More waiting—just what Doc was used to. However, it never worked. He was still producing diarrhea, had unbearable abdominal pain, and still could not eat. The pain continued to be so bad, that one particularly bad day his family rushed him to a local emergency room in the middle of the night.

The community hospital surgeon called Dr. Szema with the astonishing pathology report—over 80 eosinophils per high power field on colon biopsy! This means that when they took a sample of his colon and looked at the composition of the tissue under a microscope, it was full of inflammatory cells, known as eosinophils. You may know these cells as the mediators of allergic responses and other autoimmune conditions. The physicians were expecting to see a large amount of lymphocytes (another type of immune cell in your body) in the sample if it were truly ulcerative colitis—not tons of eosinophils!

This was not ulcerative colitis like Doc's previous physicians had speculated. This was **eosinophilic colitis**! Dr. Szema was shocked! It is common to see a few eosinophils in ulcerative colitis, but in the sample there should be many more lymphocytes. This is a very rare condition, and certainly not one that is on the top of many physician's minds. Many medical schools do not even teach about eosinophilic colitis due to the small amount of information known about it, and the small amount of the population that suffers from it.

Luckily Doc had been taken off all his medication after the colectomy, so his condition was much easier to visualize on the microscope—remember, he was previously on steroids which suppressed his immune system. Fewer eosinophils would have shown up if he was still on steroids. This would have made it even harder for other physicians to come up with an accurate diagnosis.

Although eosinophilic colitis is very rare, it is something that can be treated by downregulating IL-5 (an eosinophil activator), steroids, or also by downregulating interleukins (key factors in the allergic response). These drugs may seem familiar, because they are often also used in many allergic responses, or asthma, which also need to downregulate eosinophils.

Theoretically steroids could have helped him, but since they were not effective, other treatment options, like IL-5 inhibitors, are available for those resistant to steroid therapy.

Lucky for Doc, this was the end of his horrible journey to finding his way back to his "normal." With off label use, and approval from the hospital medical director, the head of pharmacy approved the asthma medication, anti-IL-5 receptor alpha or reslizumab, was able to take his eosinophil count from 20% to 0%. He was able to start eating again, which allowed him to gain most of the weight he had lost back and allowed him to get back to work. Although he had many trials along the way, his doctors never gave up and uncovered the answers to his mystery. This saved him many more years of pain, with the use of a single drug.

And just like that, he was able to return to his family and friends, a new man with joy back in his eyes. He was not expecting to be able to completely recover from his chronic condition; however, it was a nearly miraculous recovery! Even Dr. Szema was amazed by his impressive results. Ultimately, with his low eosinophil count (nearly 0%!), his symptoms quickly resolved, which allowed him to eat, recover his energy, and reenter society. With the proper treatment and follow-up, he was essentially able to return to life as normal once again. The continuation of the drug was the issue. Since his IL-5 inhibitor was only approved for use in asthma, this was a bit of yellow tape his medical team needed to cut through. So to get approved, Doc filed with the VA hospital for the first time. Next he and Dr. Szema sent paperwork to get the medication covered there. Thankfully, they finally approved usage of this unique drug for his eosinophilic colitis, but only within the VA by a hematologist (a specific type of physician).

Background/Salient Features of the Case

So why is it so important that we take a closer look at a rare disease like eosinophilic colitis? Well, first of all, it might not be as rare of a disease as it first appears, because physicians are so unaccustomed to seeing it, that it might get easily passed up as it did in this case. In cases that do not seem to be resolved with traditional treatment plans, it is essential for the medical team to keep looking forward to new explanations. Although there are refractory cases of ulcerative colitis that require a colectomy, this was not one of those. It is very difficult to distinguish between nearly identical presentations of symptoms, but by not labeling patients, or "anchoring" them to certain diagnoses, it is only then that physicians are able to think outside of the box and uncover the true causes of these mystery cases.

In this case, because his physicians never stopped looking for answers, he was accurately diagnosed and was able to receive a simple, one drug treatment that essentially cured him of all his symptoms. Since treatment is so readily available, and non-invasive, it is important that we take a closer look at this disease and add it to our running list of potential problems when similar patients come into our offices. If you are not a physician, then it is certainly something to bring into the conversation if you feel like you are having similar symptoms as this case—because it might not have made it onto their running diagnoses for your condition.

So let us take a closer look at how eosinophilic colitis functions.

Eosinophils are specialized white blood cells that function to protect the body against parasites, or toxins (such as seen in allergic responses). Eosinophilic colitis is seen when we do not have an underlying cause for these cells to be active. So there are no toxins or parasites that we know of, but suddenly these cells show up and start using the body's resources to fire an "allergic" reaction. This is something that we know of as a "idiopathic response"—AKA we do not know why this response is happening. Since there is nothing causing the eosinophilic response, there is nothing we can remove to make it go away. For example, if you had a splinter, we could remove the splinter and the wound would heal and the response of our body to the foreign object would eventually go away. It is notable that sometimes EC does coincide with specific food allergies—meaning it can show up in people with food allergies. Since both require the action of eosinophils this should make sense. Many of the pathways in our body overlap with one another!

The hygiene hypothesis postulates that if a person was exposed to more germs/bacteria/endotoxin as a child, then their immune system would be able to develop better and thus end up with reduced allergic responses. An example would be East German farmers with low rates of asthma living above their barns before the fall of the Berlin Wall. Since eosinophils are known to be overly sensitive in both eosinophilic colitis and allergies, it could stand to reason that eosinophilic colitis could also be subject to the germ exposure patients have in their childhood. Additionally, some individuals can react to medications, especially aspirin, which results in an allergic response. Since aspirin works systematically, it could present with similar symptoms as eosinophilic colitis.

So we have eosinophils infiltrating our colon, so why is this a problem? Well they can cause quite a few non-specific problems, including:

- Abdominal pain.

- Nausea.
- Vomiting.
- Diarrhea.
- Gastrointestinal bleeding.
- Obstruction.
- Malabsorption.
- Weight loss.
- Ascites (fluid buildup in the abdomen).

As you can imagine, having a bunch of unnecessary inflammation (this is a response from your immune system after all), and excess cells hanging out in your colon all of the time, it can really mess with your digestion and completely destroy your eating habits. Chronic inflammation is NEVER a good thing for your body to handle!

Diagnosis

To diagnose eosinophilic colitis, we first need to identify gastrointestinal symptoms, the presence of increased eosinophils in the blood, ascites, with no known causes of the eosinophilia. If there are increased eosinophils, then it is necessary to check the patient medication list to see if there could be an inciting factor, such as aspirin sensitivity. Since Doc was off of all his medications, this step was omitted from Dr. Szema's differential diagnosis. We then need to follow up with endoscopy to visualize the colitis and take a sample to view under a high powered magnifying glass. There is no defined cutoff, but if we see more than 20 eosinophils per high power field (AKA microscope picture), then it is consistent with a working diagnosis of eosinophilic colitis. It is notable that eosinophilic colitis should be consistent with eosinophils in the colon only, and not throughout the body. This is why we are going to largely see gastrointestinal symptoms.

Treatment

The current treatment for eosinophilic colitis has been largely corticosteroids—however, as you can see from this case, it was not effective for Doc and posed the risk of attendant side effects of steroids. It is our thought that more specific anti-eosinophilic medications, such as (benralizumab) would be better for eosinophilic colitis patients. Benralizumab acts to interrupt cellular signaling to eosinophils to mature. These signals, called cytokines, are used by your body in all sorts of inflammatory processes, but it is the Il-5 specific cytokines that eosinophils know to react to. By blocking the maturation of eosinophils, we can ultimately block the overload that is present in the colon of patients like Doc. No eosinophil reaction—no more symptoms! Since this blocks the single, it will not cure you from having eosinophilic colitis, but it will prevent the symptoms that follow. In other words, if you stop blocking the signal, the eosinophiles will start returning to your colon and so will your symptoms.

Fun Facts

1. Eosinophils mediate inflammatory responses such as in the context of clinical allergies and parasitic infections.
2. Eosinophilic colitis results from dysregulated enhanced recruitment of eosinophils to the colon, resulting in chronic inflammation.
3. Eosinophils are potentiated and activated by the cytokine, Il-5, so the plausibility of drugs that block IL-5 (benralizumab, mepolizumab, reslizumab) has a scientific rationale.

Questions

1. Would the eosinophil cell count be increased if Doc had a parasite infection, like tapeworms?
   (a) Yes.
   (b) No.

2. Why would an asthma medication work on a patient with eosinophilic colitis?

(a) People with asthma get eosinophilic colitis secondarily.
(b) The reactions use similar mediators and cell types.
(c) Asthma medication in particular works on a bunch of different unrelated conditions.

3. Why would a colectomy not work in a patient like Doc?
   (a) It would in a portion of patients, Ian was just unlucky.
   (b) Colectomies are curative for eosinophilic colitis, but he just needed to wait longer for it to work.
   (c) Only a portion of his colon was removed—so there were still parts that had large amounts of eosinophils built up.
   (d) The problem was not actually in his colon, but caused from his asthma.

Answers

1. (a)
2. (b)
3. (c)

Where to Find More Information:

1. Del Arco CD, et al. Eosinophilic colitis: case series and literature review. Pathol Res Pract. 2018;214(1):100–4. https://doi.org/10.1016/j.prp.2017.09.029.
2. IBD. Data and statistics 2020. https://www.cdc.gov/ibd/data-statistics.htm.
3. Barnes PJ. How corticosteroids control inflammation: quintiles prize lecture 2005. Br J Pharmacol. 2006;148(3):245–54. https://doi.org/10.1038/sj.bjp.0706736.
4. Okpara N, Aswad B, Baffy G. Eosinophilic colitis. World J Gastroenterol. 2009;15(24):2975–9. https://doi.org/10.3748/wjg.15.2975.

From Carnivore to Cautious

5

Christopher Radcliffe

C. Radcliffe (✉)
Yale University School of Medicine, New Haven, CT, USA
e-mail: christopher.radcliffe@yale.edu

A. M. Szema (ed.), *Unusual Diseases with Common Symptoms*,
https://doi.org/10.1007/978-3-031-40280-7_5

Case Presentation

Mr. Cheyenne du Boeuf, an 84-year-old gentleman, presented to the emergency department (ED) in late September for new-onset facial swelling and rash after eating sausage. This was the first time this had ever happened to him; however, he had a past history

Table 5.1 Laboratory studies for initial presentation

Laboratory study	Patient's value	Reference range	Interpretation
Serum IgE	**641 IU/mL**	<25 IU/mL	**Very high**
C1 esterase inhibitor	**16 mg/dL**	21–39 mg/dL	**Low**
C4 complement level	19 mg/dL	10–40 mg/dL	Normal
Galactose-α-1,3-galactose IgE	**44.6 kU/L**	<0.35 kU/L	**Very high**

of allergic reactions to erythromycin—an antibiotic—and poison ivy. Before this sausage incident, he had recently received treatment following exposure to poison ivy. His hobbies included gardening, and he had suspected exposure to deer as a result. To better understand the cause of his condition, physicians in the ED ordered a few laboratory studies. Table 5.1 summarizes their findings. He also had a complete blood count (CBC) with differential orders, but it was unrevealing. Typically, a CBC with differential analyzes the different types of cells in the bloodstream and is useful for detecting abnormalities caused by disease states ranging from cancer to infection.

When physicians are faced with challenging, or really even boring cases, a differential diagnosis needs to be constructed. This consists of a list of possible diagnoses that fit the clinical picture created by a patient, his/her story, physical exam findings, and laboratory studies. Given the history of facial swelling and rash following meat consumption, allergic angioedema was placed on the differential diagnosis, and the markedly elevated galactose-α-1,3-galactose IgE level supported this train of thought. His history of definite exposure to wooded areas and gardens in the Northeast U.S. put him at increased risk for arthropod bites, specifically ticks. Strangely enough, cases of new-onset meat allergies secondary to *Amblyomma* spp. (Lone star tick) bites have been reported [1], and this patient's story was deemed to be consistent with such reports. An antigen (e.g. galactose-α-1,3-galactose) is a target of sorts that the immune system can generate antibodies against, and Fig. 5.1 outlines how this process happens after a tick bite. First, a tick bites a deer and takes in the deer's blood. The

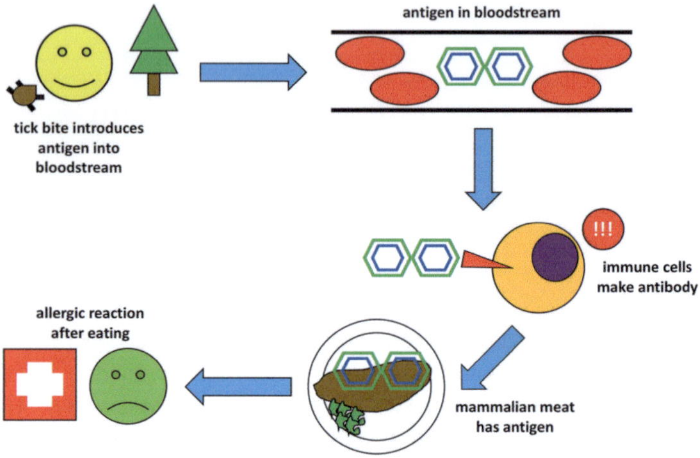

Fig. 5.1 From tick bite to mammalian meat allergy

same tick then bites a human and transfers deer carbohydrate—the antigen—during that bite. The poor human then makes allergic antibodies to the carbohydrate.

For our patient, he was eventually discharged from the ED with antihistamines and an epinephrine auto-injector prescription in case he experienced similar symptoms in the future.

Roughly 1 month after his initial evaluation, he presented to an outpatient clinic with complaints of a rash in his gluteal region. The area in question was noted to have localized redness with some evidence of hyperpigmentation. He had been avoiding red meat and milk products, so the new symptoms were unlikely to be related to his previous visit to the ED. Given his known exposures to ticks and the outdoors, Lyme disease was considered [2]. As a result, he was empirically treated with a course of doxycycline, an antibiotic. Serology labs (which check for antibodies to the bacteria responsible for Lyme disease) were subsequently negative for Lyme disease, and his C1 esterase inhibitor function test for hereditary angioedema was found to be normal (90%; ref. range > 67%). Repeat labs were drawn 10 days later and found to be normal. Altogether, the workup was negative.

He was seen at a follow-up appointment 1 month later and had been doing well. Despite the meat allergy, he had been enjoying ice cream and meatballs regularly. No episodes of angioedema, urticaria (in layman's terms—hives), or anaphylaxis (a serious allergic reaction involving multiple body systems) had occurred.

Unfortunately, in late December, he experienced an episode of urticaria accompanied by shortness of breath. Meats, sweets, and temptations abound during that time of the year, and his symptoms likely stemmed from his mammalian meat allergy. Fortunately, they resolved after taking antihistamines and using his epinephrine auto-injector. A repeat galactose-α-1,3-galactose IgE level was persistently elevated (16.10 kU/L), and his cow milk IgE level (6.30 kU/L; ref. range < 0.10 kU/L) was also elevated. Meat and milk products remained dangerous for this patient.

Diagnosis

Allergic angioedema secondary to galactose-α-1,3-galactose allergy.

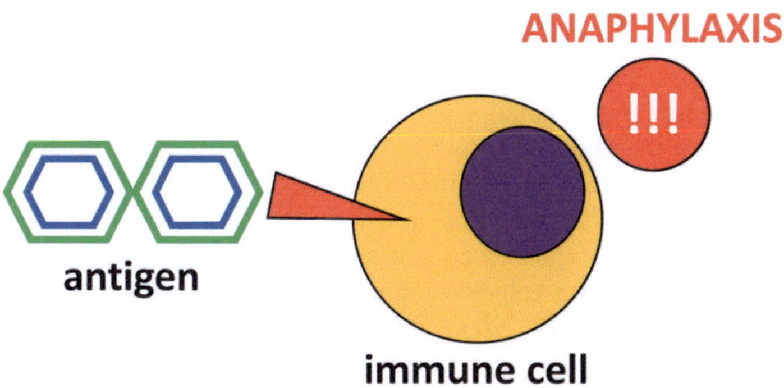

Salient Features

This case concerns an 84-year-old gentleman who initially presented with facial swelling and rash suggestive of allergic angioedema. The next question: Why did this happen? Given his history of gardening and tick exposures, he likely sustained a tick bite from a member of the *Amblyomma* spp. and developed a new-onset galactose-α-1,3-galactose allergy [1]. This allergen is found within mammalian meat products like steak, lamb, or pork. Of note, his angioedema was new. This symptom is generally defined as nonpitting edema of subcutaneous or submucosal tissue found in nearly any region of the body [3]. It may or may not be accompanied by pruritus (itching), urticaria, or anaphylaxis. Bradykinin is the main vasoactive mediator behind attacks for patients with hereditary angioedema, but general angioedema involves several proteolytic cascades [3].

The differential diagnosis for angioedema is broad and includes hereditary angioedema, allergic angioedema, angiotensin converting enzyme (ACE) inhibitor-associated angioedema, and several other types [3]. Our patient was not taking any ACE inhibitor medications (e.g. lisinopril), excluding the diagnosis of ACE inhibitor-associated angioedema. Hereditary angioedema usually presents in adolescence and rarely involves urticaria [3], so this diagnosis is also unlikely. His C1 esterase inhibitor test (level and function) was normal, excluding the diagnosis. Patients with hereditary angioedema experience several "attacks" over the course of their lifetimes, and the condition is challenging to diagnose due to its variable features. Overall, the diagnosis of allergic angioedema is favored by his newly diagnosed allergy, a past history of allergies, and urticaria during his second incident.

Galactose-α-1,3-galactose allergies have been shown to stem from lone star tick bites, although the exact mechanism is unknown [1]. Allergic symptoms usually appear a few hours after exposure to mammalian meat products and may include urticaria, angioedema, gastroenteritis (e.g. nausea and diarrhea), or anaphylaxis. The delay in time between exposure and symptom onset is thought to stem from the time delay between breakdown of galactose-α-1,3-galactose oligosaccharides (chains of sugars) in the intestinal tract and circulation of allergenic forms in the blood [1].

Interestingly, patients with mammalian meat allergies have allergic reactions to cetuximab, a monoclonal antibody used in the treatment of colorectal cancer, for it too has allergenic oligosaccharides [1, 4]. The first person to offer an explanation was Dr. Thomas Platts-Mills of the University of Virginia [5]. He first characterized the mammalian meat allergy—using himself as patient zero—and realized that wooded areas and ticks might just account for why residents of rural states are more likely to experience these phenomena relative to urbanites from NYC (Fig. 5.2). Aside from this, cetuximab is notorious for being the wonder drug containing galactose alpha 1,3 galactose that brought down Martha Stewart in an insider-trading scandal [4].

A final takeaway point from this case deals with Lyme disease, an infectious process caused by a spirochete (e.g. spiral-shaped bacteria) called *Borrelia burgdorferi* [2]. Lyme disease was first discovered in Lyme, Connecticut, and it is transmitted by the bite

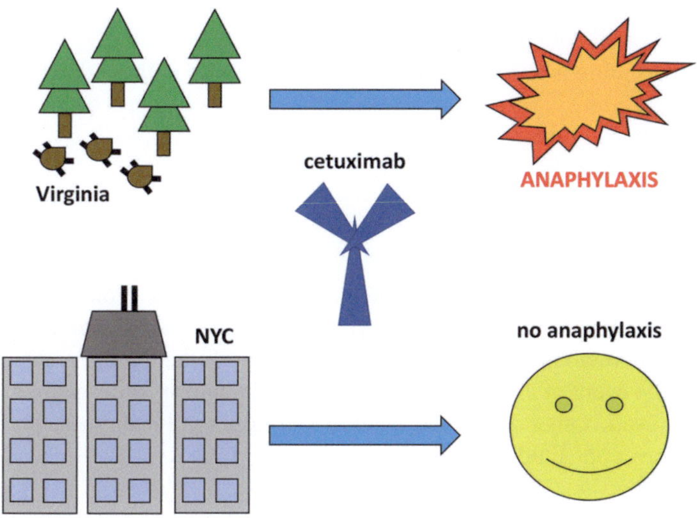

Fig. 5.2 Dr. Thomas Platts-Mills' connection

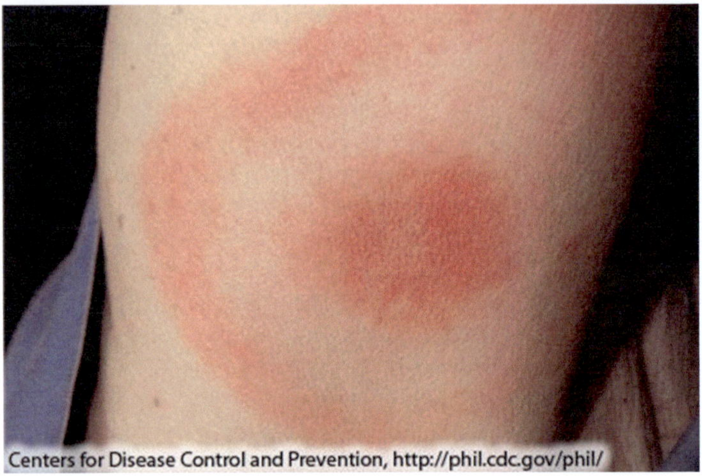

Fig. 5.3 *Erythema migrans*. (image in public domain)

of *Ixodes* spp. ticks. It leads to a variety of symptoms ranging from headache to heart block and classically causes a bullseye rash known as *erythema migrans* (Fig. 5.3).

Not all patients have this rash after being inoculated by a tick, so deciding whether or not to treat Lyme disease without diagnostic certainty is something physicians need to consider when practicing in endemic regions.

Treatment

The definitive treatment for new-onset allergies is avoidance of the inciting allergen. In this patient's case, that means avoidance of mammalian meat products (e.g. beef, pork, lamb, dairy, etc.). In the event of anaphylaxis, an epinephrine auto-injector ought to be used. Other therapies include steroids, antihistamines, and intravenous fluids [1].

Fun Facts

1. Lone star tick bites are rare causes of mammalian meat allergy 1.
2. Despite allergies to mammalian meat, one can safely eat chicken, eggs, and turkey (avian).
3. Whenever outdoors, prevention of tick bites can be accomplished with insect repellant (e.g. eucalyptus spray) and permethrin-treated clothing.

Multiple Choice Questions

1. Which of the following is an appropriate screening test for hereditary angioedema?
   (a) C1 inhibitor level.
   (b) C4 complement level.
   (c) Total serum IgE.
   (d) Serum bradykinin level.
   (e) All of the above.

2. The following antibody class is responsible for type I hypersensitivity reactions and commonly implicated with allergies:
   (a) IgA.
   (b) IgD.
   (c) IgE.
   (d) IgG.
   (e) IgM.

3. You find yourself practicing medicine in Connecticut. After noticing a bullseye rash and experiencing malaise, a patient presents to your outpatient clinic. She tells you about her hobbies of gardening and hiking. Which infectious process is most likely underway?
   (a) Leptospirosis.
   (b) Lassa fever.
   (c) Babesiosis.
   (d) Lyme disease.
   (e) Rocky mountain spotted fever.

Answers

1. (b)
2. (c)
3. (d)

References

1. Wong XL, Sebaratnam DF. Mammalian meat allergy. Int J Dermatol. 2018;57(12):1433–6.
2. Paules CI, Marston HD, Bloom ME, Fauci AS. Tick borne diseases—confronting a growing threat. N Engl J Med. 2018;379:701–3.
3. Busse PJ, Christiansen SC. Hereditary angioedema. N Engl J Med. 2020;382(12):1136–48.
4. Success story: Erbitux (NSC 632307). 2020.
5. Hester WP. Meet the doc who discovered the infamous red meat allergy spread by ticks. https://news.virginia.edu/content/meet-doc-who-discovered-infamous-red-meat-allergy-spread-ticks. Published Feb 8, 2019. Accessed 18 Apr 2020.

Watch What You Drink and Eat: Getting Orange Crushed

6

Anthony M. Szema

A. M. Szema (✉)
Columbia University Global Psychiatric Epidemiology Group CDC NIOSH U01 OH011308 "9/11 Trauma and Toxicity in Childhood: Longitudinal Health and Behavioral Outcomes", New York, NY, USA

Three Village Allergy & Asthma, PLLC, South Setauket, NY, USA

A. M. Szema (ed.), *Unusual Diseases with Common Symptoms*, https://doi.org/10.1007/978-3-031-40280-7_6

Imagine this…

You are on the porch of your grandparents' farmhouse in South Carolina; overlooking the vast wheat fields in 1975. The afternoon sun is beating down on you as you try to relax. Your clothes stick to you as they are damp with sweat from the unforgiving heat and humid environment. You are lethargic and contemplating

Table 6.1 Different products or food that may contain Balsam of Peru

Fragrances	Flavorings	Medicinal purposes
• Perfumes • Deodorants • Aftershave lotion • Sunscreens • Essential oils	• Citrus fruit peels • Vanilla • Cloves • Pimento • Nutmeg • Cinnamon • Paprika • Chocolate • Tomatoes • Wine/beer	• Oral/lip medication • Topical medicine for burn, wounds, and hemorrhoids • Dental cement • Tincture of benzoin

if you should go for a quick dip in the pond behind your grandparents' property. After a couple of minutes, and utterly tired of the heat, you say to yourself: "Why not?! This is the best idea I had all day!" But before you take that walk to the pond you decide to get a drink.

You walk to your fridge and crack open a can of orange soda. You take a swig of the soda and quickly feel your throat tightening up and collapse to the floor. You are gasping for air—hopelessly waiting for someone to come to your aid. Your spouse rushes to your side; frantically trying to call 911 and the professionals do not know what to do. The soda contained a flavoring agent known as Balsam of Peru.

Some patients can develop an allergic reaction to this substance, and you do not want it contacting your esophagus. Balsam of Peru (BOP) is a contact allergen and is used in a variety of different ways such as flavorings in our foods, to being used in fragrances, and being used for medicinal purposes. Table 6.1 shows a variety of products that may contain BOP.

Background

Balsam of Peru is a naturally occurring balsam that is derived from the bark of the Myroxylon tree, which is found in both Central and South America—specifically El Salvador. In the scientific community, this liquid is referred to as Myroxylon Pereirae (MP). It is dark brown in color and is known to be extremely viscous as well as sticky. The composition of Balsam of Peru is 60–70% cinnamon and the remaining 30%–40% are resins of unknown substances. These cinnameins are a combination of cinnamic acid, benzyl cinnamate, benzyl benzoate, benzoic acid, vanillin, and eugenol. These constituents within Balsam of Peru are also contact allergens and may lead to reactions that we see that are associated with Balsam of Peru immune reactions.

Derivatives in Food

While Balsam of Peru in of itself has mostly been discontinued in most products, there are some instances when they are still being used. Most manufacturing companies will not blatantly say Balsam of Peru on its products label; instead, there are a variety of different names that can be used. Many consumers will either not know what the meaning of it is or think it is an entirely different ingredient. The names that you may see on packages are Balsamum peruvianum, Black balsam, China oil, Honduras balsam, Indian balsam, Surinam balsam, Myroxylon pereirae klotzsch resin, Myroxylon pereirae klotzsch oil, Myrosperum pereira balsam, Toluifera pereira balsam. Most of these names will be associated with flavoring of different foods or—more commonly—with fragrance products.

As aforementioned, Balsam of Peru has mostly been discontinued in many products; however, the derivatives, or constituents, of BOP are still being used to this day. The most common derivatives that are used are vanillin, eugenol, and benzoic derivatives such as benzoates and benzoic acid. While most of the constituents of BOP are extracted and added to other food products, there are instances in which they are naturally occurring. One of those instances are tomatoes. Tomatoes contain constituents of BOP which may lead to a trigger of a reaction in an individual that is MP-sensitive. Those constituents are coniferyl alcohol and cinnamic alcohol.

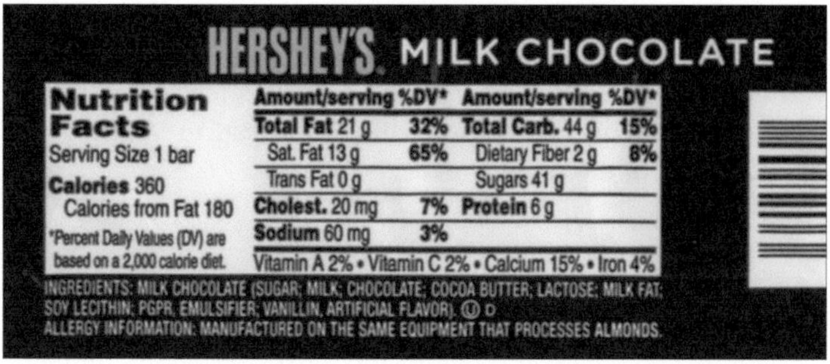

Fig. 6.1 Hershey milk chocolate nutrition facts and ingredients list

Other than tomatoes, there are constituents present in a variety of foods and beverages that we either see or have on a daily basis. One of the most popular: chocolate. Many of the chocolate brands that we see—Hershey, Ferro Rocher, Kit Kat—have the BOP constituent of Vanillin as seen in Fig. 6.1. Vanillin is described as an artificial flavoring when listed as ingredients on most products [1, 2].

Immune Response

When the skin comes into contact with allergens or irritants, we think of basic symptoms that usually occur such as the area becoming red, sore, and even itchy at the point of contact [3]. In the case of MP-sensitive patients we also may see the patient has regional swelling or inflammation. These symptoms are most commonly associated with irritant contact dermatitis. This type of dermatitis can happen to anyone; all that is needed is an irritant and chronic exposure [4]. The type of immune reaction that we see when dealing with Balsam of Peru sensitive patients is allergic contact dermatitis (ACD). This type of dermatitis is associated with delayed onset of a type IV immune reaction. This reaction is known to be a hypersensitive reaction. A person's body will react to the allergen in the most extreme and exaggerated way possible. There will be a rash and then extreme swelling [5]. Type IV reactions are characterized by cell-mediated responses which differ from the norm. The cells that are responding to the presence of antigens are T-lymphocytes. These lymphocytes will incite other white blood cells to rush to the point of contact in order to coordinate a huge immune response which will lead to swelling and inflammation of the contact area.

Conclusion

Balsam of Peru can cause one of the most exaggerated immune responses and may be fatal depending on where the contact area of the allergen is. While Balsam of Peru is not used as a whole in today's society, there are still many of its constituents that are being used in many products today. These constituents may lead to allergic reactions that we are used to see in MP-sensitive patients. Doctors may be able to conduct patch testing to see which constituent specifically elicits a hypersensitive immune response. Through this, they may be able to give their patients a more accurate way of avoiding these contact allergens.

Fun Facts

1. Balsam of Peru is a flavoring agent in foods.
2. Balsam of Peru is not only used for its aromatic and fixative (i.e. delays evaporation) properties but also for its mild antiseptic, antifungal, and antiparasitic attributes.
3. Balsam of Peru has 3 main uses: **fragrance in perfumes and toiletries; flavoring in food and drink; healing properties in medicinal products**.

Question

Is Balsam of Peru still in Orange Crush Soda?

Answer

No, but it used to be. A patient with a positive patch test to Balsam of Peru may have had throat irritation with imbibing beverages containing it.

References

1. Scheman A, Rakowski EM, Chou V, Chhatriwala A, Ross J, Jacob SE. Balsam of Peru: past and future. Dermatitis. 2013;24(4):153–60. https://doi.org/10.1097/DER.0b013e31828afab2; PMID: 23857009.
2. Katta R, Schlichte M. Diet and dermatitis: food triggers. J Clin Aesthet Dermatol. 2014;7(3):30–6; PMID: 24688624; PMCID: PMC3970830.
3. Guarneri F, Corazza M, Stingeni L, Patruno C, Napolitano M, Pigatto PDM, Gallo R, Cristaudo A, Romita P, Offidani A, Schena D, Milanesi N, Micali G, Zucca M, Foti C, SIDAPA Study Group. Myroxylon pereirae (balsam of Peru): still worth testing? Contact Dermatitis. 2021;85(3):269–73. https://doi.org/10.1111/cod.13839; Epub 2021 May 3. PMID: 33748955; PMCID: PMC8453940.
4. Atwater AR, Bembry R, Liu B, Warshaw EM, DeKoven JG, Silverberg JI, Taylor JS, Zug KA, Houle MC, Belsito DV, Maibach HI, Reeder MJ, Sasseville D, Dunnick C, Fowler JF Jr, Pratt MD, DeLeo VA. Medical adhesive allergens: retrospective analysis of cross-sectional data from the North American contact dermatitis group, 2001–2018. J Am Acad Dermatol. 2021;S0190-9622(21):02927–3. https://doi.org/10.1016/j.jaad.2021.11.055; Epub ahead of print. PMID: 34875303.
5. Pham CT, Juhasz M, Lin J, Hashemi K, Honari G, Mesinkovska NA. Allergic contact dermatitis of the scalp associated with scalp applied products: a systematic review of topical allergens. Dermatitis. 2022;33(4):235–48. https://doi.org/10.1097/DER.0000000000000844; Epub 2022 Mar 22. PMID: 35318978.

I Can Finally Light a Candle

7

Ashwin Verghese

A. Verghese (✉)
Tufts University School of Medicine, Boston, MA, USA
e-mail: ashwin.verghese@tufts.edu

A. M. Szema (ed.), *Unusual Diseases with Common Symptoms*,
https://doi.org/10.1007/978-3-031-40280-7_7

Vignette

At age 14, I visited my optometrist for my annual eye exam to obtain a new prescription for reduced vision. I had worn corrective lenses for myopia (nearsightedness) and astigmatism (irregular shape of the cornea, the clear structure at the front of the eye that is responsible for two-thirds of the eye's optical power) since second grade. However, I had also recently found that my vision was asymmetric; the visual acuity through my right eye was significantly worse than that through my left. When attempting to light a candle, I found myself missing multiple times, suggesting that my depth perception was severely weakened. I also noticed that when I attempted to view small or bright objects, such as car headlights, with my right eye, I instead perceived several distorted versions of them in a circle centered around the true position of the object.

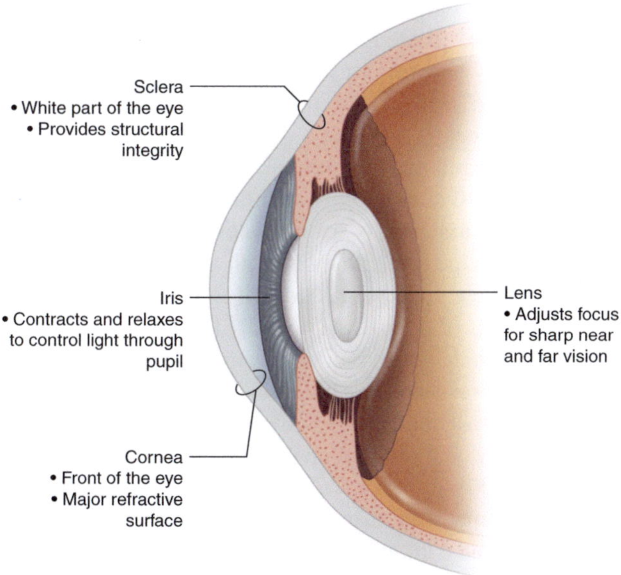

Despite corrective glasses, my vision was never correctable to 20/20. I had never had any other eye conditions, and I had never sustained any ocular injuries. I do, however, have seasonal grass and oak tree pollen allergies, and exposure may cause rashes, as well as tearing up and redness in the eye. With regard to family history, my father wears eyeglasses for myopia and astigmatism, though with no asymmetry; my mother had only begun wearing reading glasses within the previous year. My younger brother also wears eyeglasses for myopia and was later diagnosed with and treated for the same condition as me, though at an earlier stage. Additionally, various relatives have unknown conditions with low visual acuity and distorted images.

Diagnosis

At my optometrist's office, I underwent standard Snellen eye chart and slit-lamp microscopy exams. In the Snellen exam, I was found to be unable to see the large E (20/200) while wearing corrective eyeglasses with my right eye; the actual findings returned 20/125 and 20/500 in the left and right eyes, respectively. Under the slit-lamp, my optometrist found corneal scarring in my right eye. My optometrist suspected that I had keratoconus, a condition in which the cornea (clear tissue on the front of the eye) bulges outward. He referred me to a corneal ophthalmologist for corneal topography scans that would help confirm the diagnosis.

On the corneal topography scan, my corneas were extremely thin and irregularly shaped in both eyes, with the thinnest and steepest areas slightly below the pupils. A normal cornea has uniform curvature and thickness of approximately 43D (diopters, units of curvature such that a circle with a radius of ½ m has constant curvature of 2D) and 550 μm respectively (Fig. 7.1) [1]. In my scans, the maximum curvature was found to be 68.4D and 88.7D in the left and right eyes, respectively, and the minimal corneal thickness was found to be 421 μm and 349 μm (Fig. 7.2). Thus, my ophthalmologist confirmed my optometrist's suspicion of keratoconus, supported by more curvature and a thinner cornea.

See Fig. 7.2a and b.

Keratoconus is a progressive corneal disorder characterized by an abnormal cornea. The cornea may show thinning, and irregular astigmatism (bulging), and in severe cases, scarring, and may result in myopia (nearsightedness) and image distortion [1]. It normally begins during the second decade in affected individuals, though it is difficult to detect in early stages, so the patient often receives no treatment apart from that for myopia and regular astigmatism [2]. In more advanced keratoconus, the astigmatism becomes more irregular, resulting in a cone-shaped protrusion below the pupil and results in low and distorted vision that cannot be fully corrected through corrective lenses [2].

Keratoconus can result from both genetic and environmental influences. Environmental factors include the wearing of contact lenses and eye rubbing, which may also lead to increased occur-

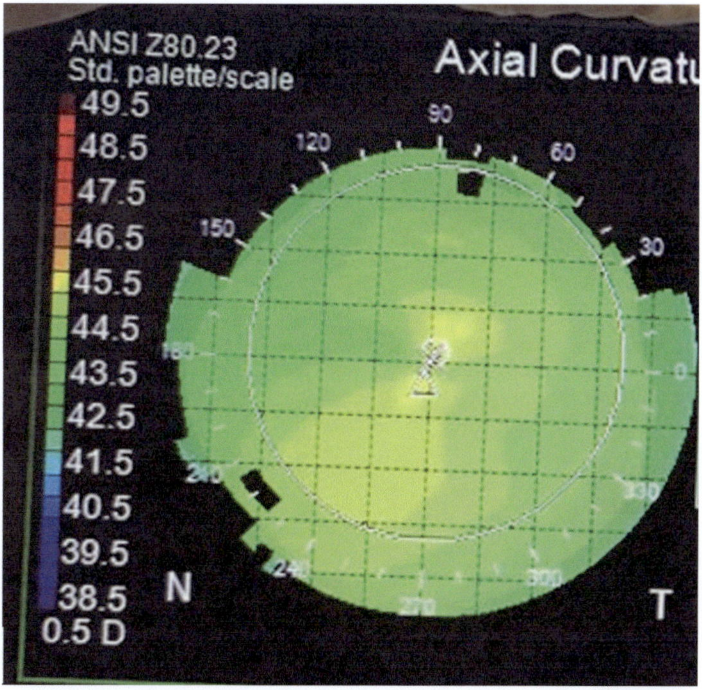

Fig. 7.1 This shows the curvature map of a patient with a relatively normal cornea. It is to be noted that the curvature is relatively constant throughout the eye near 45D (fully green). There is slight astigmatism at 0.7D demonstrated by the yellow hourglass shape, but this cornea is not keratoconic. (Provided by Ronald F. Luxenburg, OD)

rence of conjunctivitis [2–5]. Correlational studies on the heritability of keratoconus indicate an autosomal dominant inheritance pattern; despite several candidate genes, however, the genetic causation has not been clinically determined [4]. I believe my risk factors for keratoconus entail:

1. Genetic susceptibility as I have relatives with unknown eye conditions characterized by low visual acuity and distorted image production, and
2. Eye rubbing, especially in response to redness and tear production caused by allergies [5].

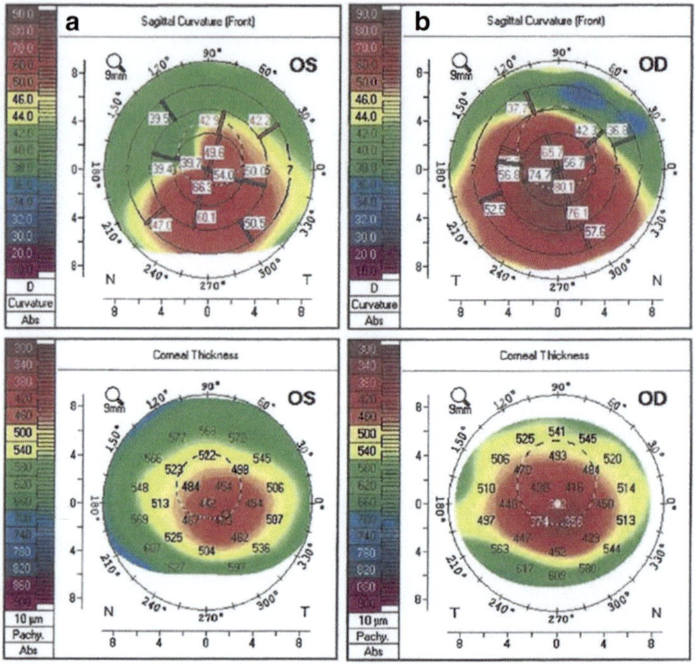

Fig. 7.2 **(top):** My diagnostic corneal topography scans in the left eye for curvature (top) and thickness; **(bottom):** My diagnostic corneal topography scans in the right eye for curvature (top) and thickness (bottom). The significant red in the curvature scans demonstrates steepening of the corneas slightly below the center of the pupils. The corneas are also extremely thin, as indicated by the large red and yellow areas in the thickness maps. (Provided by Martin L. Fox, MD, FACS)

Although early stages of keratoconus can be treated with corrective lenses, advanced keratoconus requires surgical intervention. Treatment options can include collagen cross-linking (CXL), wherein ultraviolet light is used to create cross-links between collagen subunits that make up the cornea in order to prevent further progression; implantation of intrastromal ring segments (INTACS®), which are semicircular plastic half-rings inserted to flatten the cornea in hopes of improving vision; and, in the most severe cases, corneal transplantation [4, 6]. Either deep anterior lamellar keratoplasty (DALK), which is a partial-depth corneal transplant, or penetrative keratoplasty, which is a full-depth corneal transplant, may be performed [4, 6].

Treatment

My ophthalmologist, my parents, and I decided on a two-step plan for my keratoconus treatment. In order to limit the progression of keratoconus, I first underwent cross-linking (CXL) in both eyes. Afterward, I was also implanted with ring segments in the right eye in hopes of regaining functional vision.

In preparation for the surgeries, I was prescribed gatifloxacin (one drop each at breakfast, lunch, dinner, and bedtime) for 1 week preceding the surgeries. During the CXL procedure, I was first given two drops of riboflavin in each eye every 2 min for a half-hour. Afterward, under topical lidocaine for anesthesia, the epithelium was removed, and ultraviolet light (370 nm, 3 mW/ cm^2) was applied for 5 min in both eyes [2]. During the INTACS® procedure, a stromal tunnel was created using the 60 kHz IntraLase femtosecond system, into which the semicircular ring segments were inserted [7]. Afterward, a soft contact lens was inserted in order to prevent friction between the lids and the cornea.

Postoperatively, I was required to wear shields on the eyes while sleeping for 1 week in order to prevent external pressure on the eyes during the healing process. I was prescribed gatifloxacin, an anti-inflammatory drug to prevent conjunctivitis (pink-eye), four times daily; ketorolac tromethamine, a nonsteroidal anti-inflammatory drug for management of inflammation and pain, two drops daily; and pregabalin to manage pain as needed, with a maximum of four 75 mg tablets a day. This continued for 4 weeks. I was also given prednisolone acetate (one drop each at breakfast, lunch, dinner, and bedtime), a corticosteroid to prevent scarring during the epithelial healing process, for 1 week, which was replaced afterward with fluorometholone, a weaker corticosteroid, for 3 months, starting at four times a day and reducing to three times a day after 1 month and once a day after 2 months.

See Fig. 7.3a and b.

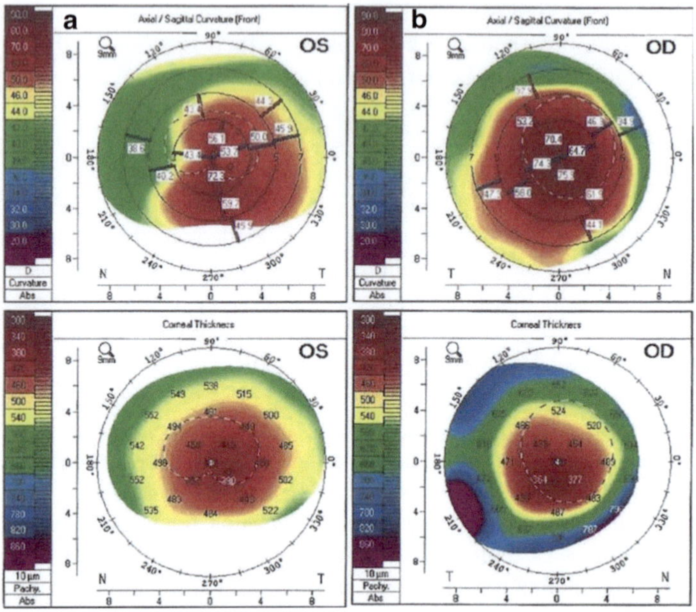

Fig. 7.3 **(left):** My postoperative corneal topography scans in the left eye for curvature (top) and thickness (bottom); **(right):** My postoperative corneal topography scans in the right eye for curvature (top) and thickness (bottom). Compared to the diagnostic scans, the maximum curvature is greater in the left eye at nearly 73D, and there are larger areas of red, indicating thinning. This may be due to advancement between the date of diagnosis and the surgeries, which happened for 9 months. (Provided by Martin L. Fox, MD, FACS)

There were no complications or side-effects from the surgeries (Fig. 7.3). After my eyes healed completely, and no progression was determined after 6 months, my ophthalmologist recommended that I speak with my optometrist about contact lenses, as he expected that it would provide better corrected vision than glasses. My optometrist fitted me with scleral lenses, which are large lenses that extend partially over the sclera (white part of the eye). Five years after the operation, my corneal thickness and curvature was found to be stable, and my corrected vision in both eyes was 20/25, though slightly worse in my right eye (Fig. 7.4).

See Fig. 7.4a and b.

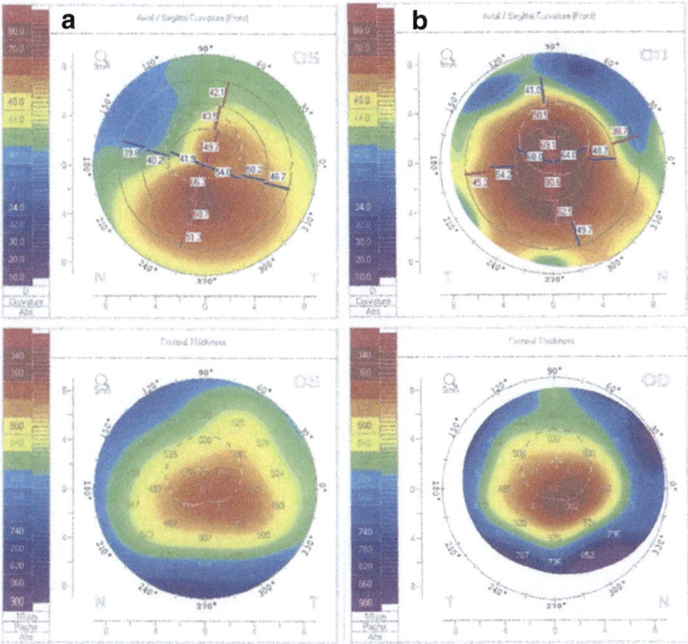

Fig. 7.4 **(top):** My 5-year postoperative corneal topography scans in the left eye for curvature (top) and thickness (bottom); **(bottom):** My 5-year postoperative corneal topography scans in the right eye for curvature (top) and thickness (bottom). There appears to be no major differences between the 5-year and the immediate postoperative scans, indicating success in halting the progression of the disorder. (Provided by Martin L. Fox, MD, FACS)

Discussion

Keratoconus is a progressive corneal degenerative disease that often presents at puberty and is characterized by the thinning and steepening of the cornea due to the change in the structure of collagen, the protein that composes it [3, 5]. One of the factors that increases the likelihood of developing keratoconus, and increases its rate of progression, is eye rubbing [3]. Additionally, several studies have found a high prevalence of skin allergy, allergic conjunctivitis, and asthma in keratoconus patients [3]. Patients with vernal keratoconjunctivitis have been shown to have earlier onset of keratoconus with more severe thinning and also show increased intraocular pressure (pressure exerted on the eye by fluid produced within) [5]. Allergy has also been found to increase the risk of hydrops (leakage of fluid from within the eye) [3].

Early stages of keratoconus may not be distinguishable from general myopia and regular astigmatism; however, corneal topography is useful as it can be used to detect mild keratoconus [3]. After my surgeries, my younger brother, then age 11, had his cornea scanned, even though he had no other signs indicating keratoconus as well. He was found to have keratoconus as well. He received CXL in both eyes to halt its progression but did not need further treatment. Thus, it can be recommended that patients known to be at risk for keratoconus and astigmatic adolescents with a history of allergy to be screened during routine eye exams with corneal topography scans [3]. This can be helpful in arresting the progression of keratoconus early on and preventing severe visual acuity loss and image distortion.

Fun Facts

1. Keratoconus may present with myopia and astigmatism that cannot be corrected with eyeglasses.
2. Corneal scarring may be observed under the slit lamp; diagnosis of keratoconus can be made through corneal topography.
3. Treatment options include scleral contact lenses, CXL intrastromal ring segment implants, and corneal transplantation.

Questions

1. Which of the following statements about keratoconus is INCORRECT?
   (a) Keratoconus is primarily a genetic disorder.
   (b) Keratoconus causes myopia, cone-shaped irregular astigmatism, and scarring.
   (c) Early stages of keratoconus are often treated similarly to myopia and astigmatism.

(d) Patients with advanced keratoconus may need corneal transplants.
(e) The cone-shaped protrusion often occurs inferior to the pupil.

2. Which of the following treatments is NOT commonly used to treat keratoconus?
   (a) Corneal collagen cross-linking (CXL).
   (b) Penetrating keratoplasty (PK).
   (c) Intrastromal ring segments (INTACS®).
   (d) Deep anterior lamellar keratoplasty (DALK).
   (e) Laser-Assisted In Situ Keratomileusis (LASIK).

3. Which of the following is matched with its CORRECT purpose?
   (a) Eye shield; prevention of ocular trauma.
   (b) Prednisolone acetate; pain management.
   (c) Ketorolac tromethamine; prevention of scarring.
   (d) Pregabalin; anti-inflammatory drug.
   (e) Gatifloxacin; pain management.

Answers
1. (a)
2. (e)
3. (a)

References

1. Iyamu E, Osuobeni E. Age, gender, corneal diameter, corneal curvature and central corneal thickness in Nigerians with normal intraocular pressure. J Optom. 2012;5(2):87–97. https://doi.org/10.1016/j.optom.2012.02.001.
2. Maier P, Reinhard T, Kohlhaas M. Corneal collagen cross-linking in the stabilization of keratoconus. Dtsch Arztebl Int. 2019;116(11):184–90. https://doi.org/10.3238/arztebl.2019.0184.
3. Sharma N, Rao K, Maharana PK, Vajpayee RB. Ocular allergy and keratoconus. Indian J Ophthalmol. 2013;61(8):407–9. https://doi.org/10.4103/0301-4738.116063.

4. Davidson AE, Hayes S, Hardcastle AJ, Tuft SJ. The pathogenesis of keratoconus. Eye. 2014;28(2):189–95. https://doi.org/10.1038/eye.2013.278.
5. Cingu AK, Cinar Y, Turkcu FM, et al. Effects of vernal and allergic conjunctivitis on severity of keratoconus. Int J Ophthalmol. 2013;6(3):370–4. https://doi.org/10.3980/j.issn.2222-3959.2013.03.21.
6. Parker JS, van Dijk K, Melles GRJ. Treatment options for advanced keratoconus: a review. Surv Ophthalmol. 2015;60(5):459–80. https://doi.org/10.1016/j.survophthal.2015.02.004.
7. Alio JL, Vega-Estrada A, Esperanza S, Barraquer RI, Teus MA, Murta J. Intrastromal corneal ring segments: how successful is the surgical treatment of keratoconus? Middle East Afr J Ophthalmol. 2014;21(1):3–9. https://doi.org/10.4103/0974-9233.124076.

A Candyman's Popcorn Lungs

8

Timothy C. Olsen
and Robert A. Promisloff

Vignette
Ten-feet, 20-feet, 30-feet high walls laden with colorfully-sugared assortments, punctuated with sweet aromas, embody candy-filled fantasies of candy-filled factories. For Caramelo Golosina, such factories lured him into its dizzying spell as a candyman after high school. His days meant stirring pots of bubbling flavors, concomitantly inhaling flooding vapors (Fig. 8.1).

T. C. Olsen (✉)
University of Rochester School of Medicine & Dentistry, Rochester, NY, USA
e-mail: timothy_olsen@urmc.rochester.edu

R. A. Promisloff
Drexel University College of Medicine, Philadelphia, PA, USA

A. M. Szema (ed.), *Unusual Diseases with Common Symptoms*,
https://doi.org/10.1007/978-3-031-40280-7_8

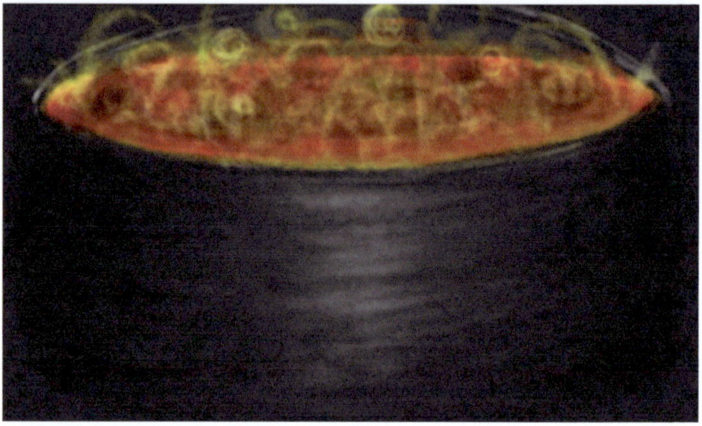

Fig. 8.1 Boiling candy

https://schneiderinvestigates.wordpress.com/category/diacetyl/

After several years of employment, 25-year-old Caramelo presented to our pulmonary office complaining of shortness of breath, after walking as little as one block or a flight of stairs, with occasional wheezing. These symptoms had been present for approximately 3 months. He had been well, with no history of pulmonary or cardiac diseases, without asthma, and he was not on regular medications. He was a life-long city resident of North Philadelphia employed as a candy factory worker.

Mr. Golosina's physical examination was unremarkable except for expiratory wheezing heard as a high-pitched moan akin to an orca crying. Medically, expiratory wheezing results from a narrowing of the lower airways, common in asthma and tobacco-related pulmonary illnesses, like Chronci Obstructive Pulmonary Disease (COPD).

A pulmonary function test (PFT), a quantitative evaluation of the patient's breathing, revealed severe obstructive airways disease. Simply put, the patient could inhale but could not exhale normally due to narrowed airways (Fig. 8.2). His PFT showed that he could only expire 40% of his air in 1 second, compared to the normal 70–80%. There was no improvement after inhalation of a bronchodilator (albuterol). Additionally, his elevated residual volume of 125% of predicted indicated air trapping (air can enter, but not exit) (Fig. 8.2). Hyperinflation seen on the chest X-rays was consistent with air trapping (Fig. 8.3).

After an extensive workup, which had failed to yield a diagnosis, or improvement with standard asthma treatments, a lung biopsy was conducted. Surprisingly, the biopsy revealed significant luminal narrowing, swelling, and scarring of the small airways, indicating an inflammatory and fibrotic small airway disease (Figs. 8.4 and 8.5). Caramelo's symptoms, together with his PFT results, radiographs, and lung biopsy were consistent with the unexpected diagnosis of obliterative bronchiolitis—the question of how and why, nonetheless, remained.

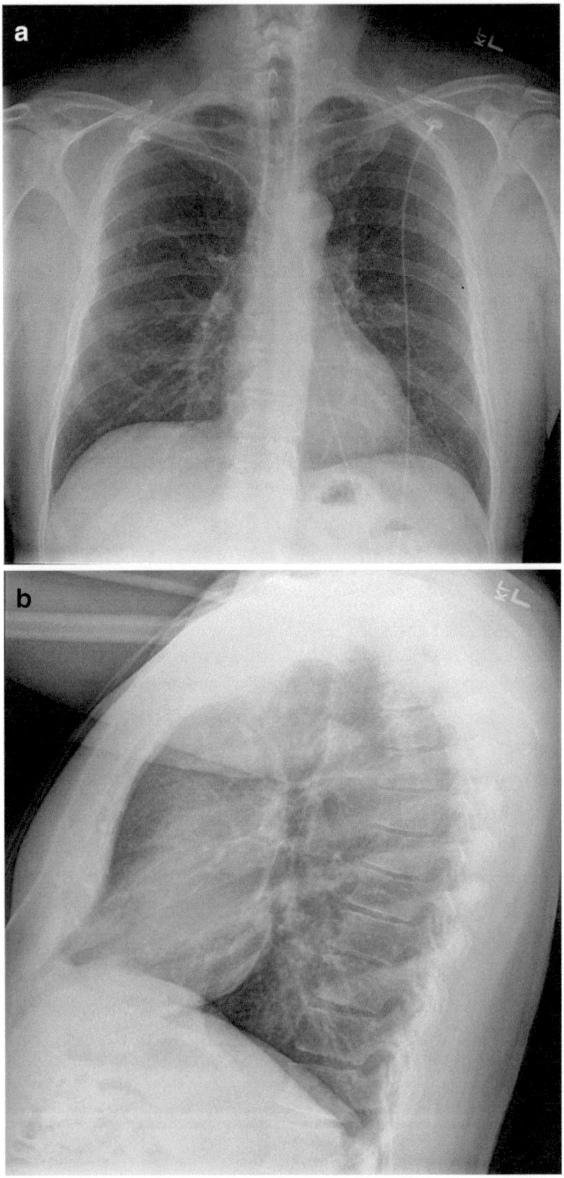

Fig. 8.2 (**a**) Posterior–anterior chest X-ray, showing hyperinflation indicated by mild diaphragm flattening in another patient with obliterative bronchiolitis (OB) [1]. (**b**) Lateral chest X-ray, showing hyperinflation indicated by mild diaphragm flattening in an OB patient [2]

	Patient's Results	Normal for Demographic Profile
Forced Vital Capacity (FVC)	65% Predicted	≥80% Predicted
Forced Expiratory Volume in 1 Second (FEV1)	35% Predicted	≥80% Predicted
FEV1/FVC	40%	>70%
Total Lung Capacity (TLC)	110% Predicted	80-120% Predicted
Reserve Volume (RV)	125% Predicted	80-120% Predicted
Diffusing Capacity for Carbon Monoxide (DLCO)	80% Predicted	≥80% Predicted

Fig. 8.3 Pulmonary function test (PFT) results

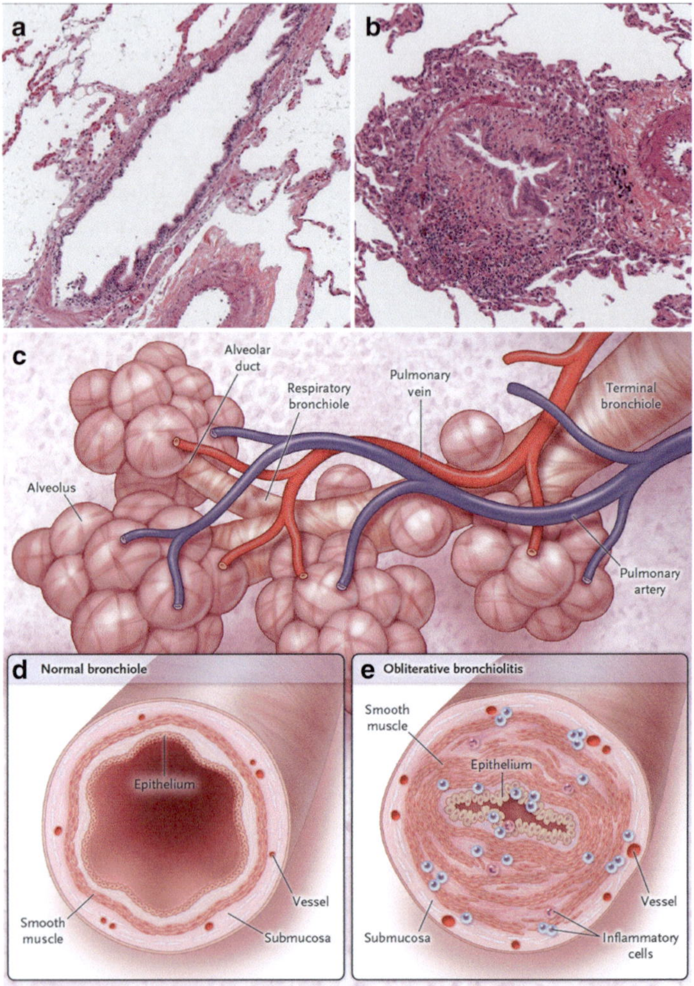

Fig. 8.4 The top left and right sections are distal bronchiole sections: normal bronchiole (**a**) vs. obliterative bronchiolitis affected bronchiole (**b**). The middle image (**c**) illustrates the anatomy of the lower respiratory system. The terminal bronchioles lead to the alveol or air sacs, where, inside, gas-exchange occurs. The bottom left and right images are cross-sections of a normal bronchiole (**d**) and a diseased bronchiole (**e**). Notice the narrowing and constrictive properties caused by obliterative bronchiolitis (**e**) [3]

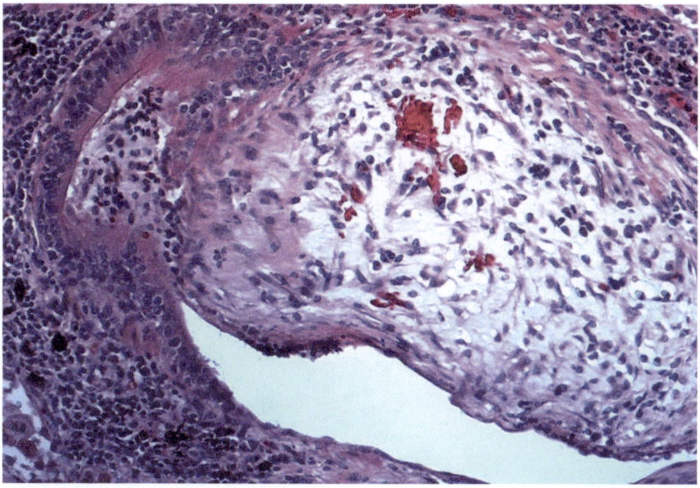

Fig. 8.5 Lung biopsy of a distal bronchiole illustrating constriction of airway, growth of polypoid nodule of immature connective tissue, and fibroblasts [4]

Background/Salient Features of the Case

To understand the etiology behind Mr. Golosina's case, one must analyze his clinical presentation. Surprisingly, the least relevant fact that he self-disclosed was critical—he worked in a candy factory.

Caramelo's likely exposure to an artificial flavoring agent, diacetyl, plausibly incited his inflammatory lung disease while working in a North Philadelphia candy factory. Diacetyl is commonly used to impart a buttery flavor to foods, such as butterscotch and toffee flavored candy, and most widely, popcorn. Toxicity stems from the diacetyl vapors escaping its yellow-green bath and entering the victims' lungs (not wearing a mask). Once inside, the chemical agent can form harmful connections with cells, breaking down proteins, and even untangling DNA (Fig. 8.6). Over time, the vapors can accumulate and irreversibly inflame the small airway bronchioles. In turn, chemically prompt-

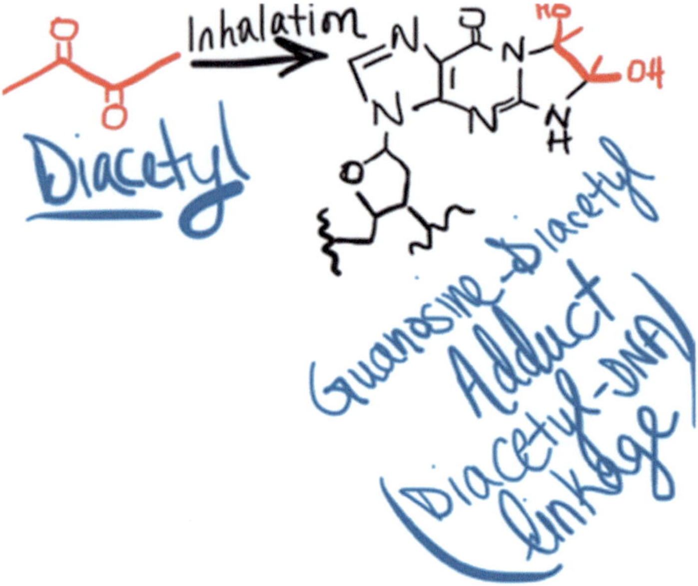

Fig. 8.6 One of diacetyl's (red) mechanisms of toxicity: diacetyl can link with guanosine, a fundamental building block of DNA

ing the lining cells to secrete mucus abundantly and attracting white blood cells further feeds lung inflammation (Fig. 8.4). Mucus plugs consisting of dead cells, pus, and scar tissue can then form and create de facto one-way valves, trapping air, and disrupting oxygen-blood diffusion. Caramelo's wheezing and obstructive lung function are thus pathologically explainable and shared among other patients similarly plagued with obliterative bronchiolitis.

Other fumes that cause comparable airway scarring and narrowing, while rare, include nitrogen oxides used in fertilizer production, sulfur mustard gas used post-Iran-Iraq War (1980–1988), and sulfur dioxide gas used in the Iran-Iraq War and WWI (1914–1918). In urban and low socioeconomic strata areas, sulfur dioxide and nitrogen oxides are common pollutants from automobile exhaust and fossil fuel-powered factories. In the atmosphere, these oxides can form acid rain and increase the infectivity of viruses.

In infants under 2 years of age, lower airway fibrosis and inflammation can commonly result from RSV, influenza, adenovirus, measles, and mycoplasma. Infant bronchiolitis can become so severe that lips, fingers, and toes eerily become cyanotic-blue from a lack of oxygen. In older individuals, obliterative bronchiolitis commonly follows lung transplants, hematopoietic stem cell transplants (HSCT), and connective tissue disorders, such as rheumatoid arthritis. The believed cause for transplant patients is either host rejection of the donated organs (lung transplant patients) or an overreaction of the received donor stem cells against native lungs (HSCT patients). In rheumatoid arthritis, an autoimmune disorder that damages joints, inflammatory damage can also be directed against the lungs' small airways, hence bronchiolitis.

Our patient is unique by not fitting these demographics or medical histories; yet, this is not an entirely isolated case. From 1993 to 2000, eight workers at a Missouri popcorn factory began complaining of increasing difficulty breathing, lethargy, dry cough, and night sweats. They saw varying physicians, including allergists and pulmonologists. Eventually, in May of 2000, the eight cases were reported as severe obliterative bronchiolitis to the Missouri Department of Health, bearing investigations into the popcorn factory. As a result, the investigations found diacetyl as the likely cause [5].

Furthermore, the workers responsible for mixing the 130 °F (54 °C) butter-substitute solutions, vaporized and inhaled, were most susceptible—a position similar to Mr. Golosina's butterscotch mixing post. News of this disease caught the attention of the media and nation, colloquially becoming known as popcorn lung and popcorn workers' lung. In response, many food industry factories have adopted new ventilation precautions and have phased out diacetyl from their ingredients for workers' safety [5]. However, many e-cigarette companies are still utilizing the causative agent in their liquid cartridges—potentially posing a threat to their production workers and their customers who purposely inhale the vaporized liquids (Figs. 8.7 and 8.8).

Fig. 8.7 Popcorn [6]

Fig. 8.8 Butterscotch candy [7]

Diagnosis

In Mr. Golosina's case, he presented as a young, previously-healthy individual with severe shortness of breath on exertion and severe obstruction on PFTs. There was no history of pulmonary or cardiac disease, nor of tobacco or drug use. Additionally, there was no immediate family history of a hereditary cause of emphysema, alpha1-antitrypsin deficiency. In this 25 year old's age group, asthma would be the most common cause of shortness of breath, wheezing, and obstructive PFTs. However, this patient had no previous history of asthma or allergies, nor markers for either. Only until the analysis of Caramelo's lung biopsy together with his previous tests was his rare diagnosis of obliterative bronchiolitis confirmed.

Treatment/Conclusion

Treatment for obliterative bronchiolitis is challenging due to the irreversible accumulation of fibrotic scar tissue in the airways. As a result, the aim of its treatment often focuses on decreasing disease progression. It should be apropos to the cause and person. Medications range from immunosuppressants to macrolide antibiotics. Such antibiotics, like azithromycin, interestingly, serve not only the typical function against bacterial infections but also the concomitant one in reducing neutrophil-induced inflammation, i.e., self-damage from the immune system.

As noted earlier, in transplant patients, rejection serves as a significant cause of bronchiolitis. Drugs that help transplant patients reduce or prevent bronchiolitis from organ rejection include immunosuppressants, such as tacrolimus. By suppressing the immune system, tacrolimus can reduce airway inflammation and hyperactive immune responses toward the host's lungs.

Additionally, inhaled bronchodilators, corticosteroids, and leukotriene antagonists are other treatment options that can assist airway patency. There is no one-size-fits-all treatment for this disorder and care may often require a long-term combined therapeutic approach.

Caramelo was followed-up for 18 years after his initial presentation, treated with a combination of bronchodilators, inhaled corticosteroids, and leukotriene antagonists. Exposure to toxins from his workplace also ceased immediately. Over these 18 years, his clinical status and lung function remained stable but strikingly low.

While there is no definitive treatment, stopping exposure at the workplace was the main factor in Caramelo living a relatively stable life. His case highlights the importance of not passively accepting common diagnoses like asthma at the sound of wheezing and is why our one time candymaker remains stable today. An apt quote made in the 1930s by physician Chevalier Jackson indeed rings true today, "all that wheezes is not asthma."

Fun Facts

- Diacetyl can be sourced both naturally as a byproduct of fermentation (found in dairy products like butter, cheese, milk, bread, coffee, brandy, and rum) and artificially (found in candy, baked goods, cake mixes, and butter-flavoring for popcorn) [8].

❷ No cases of OB have been reported in consumers of butterscotch candy.

❸ R. Miler et al. reported a staggering 38 out of 49 US soldiers as having constrictive bronchiolitis (obliterative bronchiolitis), plausibly from the inhalation of toxic aerosols released from a 2003 sulfur-mine fire in Iraq and from burn pits while also deployed in Afghanistan [9].

Multiple Choice Questions

1. Which statement is false?
   (a) Obliterative bronchiolitis is a distinct pattern of small airway disease characterized by subepithelial inflammation and narrowing of the bronchioles.
   (b) The most common cause of expiratory wheezing in adults is obliterative bronchiolitis.
   (c) Treatment is different for asthma and bronchiolitis.
   (d) Obliterative bronchiolitis most commonly presents in infants under 2 years of age.
   (e) Diacetyl vapors, sulfur mustard gas, and RSV can cause obliterative bronchiolitis.

2. What is/are diacetyl's mechanisms of action vis-à-vis obliterative bronchiolitis?
   (a) Diacetyl acts as a nucleophile against guanosine, binds, and initiates untangling of DNA.
   (b) Arginine's guanidino group acts as a nucleophile against diacetyl, forming arginine-diacetyl adducts among proteins found in the respiratory tract's epithelial and subepithelial cells.
   (c) Diacetyl acts as an electrophile against arginine's aliphatic 3 carbon chain, binds, and disrupts structures and functions of proteins lining the respiratory tract.
   (d) Choices (a) and (b) are correct.
   (e) Choices (b) and (c) are correct.

3. How is obliterative bronchiolitis definitively diagnosed?
   (a) An FVC ≤79%
   (b) An FEV1 ≤79%
   (c) A radiograph image indicating hyperinflation, seen in the flattening of the diaphragm.
   (d) A distal airway biopsy indicating constriction and fibrosis.
   (e) Either (a), (b), or (c).

Answers

1. (b)
   The most common cause of wheezing in adults is obliterative bronchiolitis.
2. (b)
   Note: (a) Diacetyl acts as an *electrophile* against guanosine. (b) Diacetyl acts as an electrophile against arginine's *guanidino* group.
3. (d)
   Note: while (a), (b), and (c) are likely to present in obliterative bronchiolitis, they are neither necessary nor sufficient criteria for the diagnosis of OB. Oftentimes soldiers present with technically normal PFT and radiographic images, but when biopsied, reveal fibrotic constriction of the lower airways [9].

References

1. By O'Dea—own work, CC BY-SA 3.0. https://commons.wikimedia.org/w/index.php?curid=28065676.
2. By O'Dea—own work, CC BY-SA 3.0. https://commons.wikimedia.org/w/index.php?curid=28065675.
3. Barker AF, Bergeron A, Rom WN, Hertz MI. Obliterative bronchiolitis. N Engl J Med. 2014;370(19):1820–8. https://doi.org/10.1056/nejmra1204664.
4. By Yale Rosen from USA—cryptogenic organizing pneumonia (COP)-bronchiolitis obliterans. Uploaded by CFCF, CC BY-SA 2.0. https://commons.wikimedia.org/w/index.php?curid=31128019.
5. Kreiss K, et al. Clinical bronchiolitis obliterans in workers at a microwave-popcorn plant. N Engl J Med. 2002;347(5):330–8. https://doi.org/10.1056/nejmoa020300.

6. Maxwell R. File:Popcorn (2838177608).jpg [September 7 237/366]. 2008. https://commons.wikimedia.org/wiki/File:Popcorn_(2838177608).jpg. Accessed 2 Sep 2020.
7. Evan-Amos. File:Butterscotch-Candies.jpg [Butterscotch hard candies. Commonly sold in America, these are penny candies wrapped in a nondescript yellow cellophane wrapper]. 2012. https://upload.wikimedia.org/wikipedia/commons/6/6d/Butterscotch-Candies.jpg. Accessed 2 Sep 2020.
8. Insight F. Questions and answers about diacetyl. 2018. https://foodinsight.org/questions-and-answers-about-diacetyl/.
9. King MS, Eisenberg R, Newman JH, Tolle JJ, Harrell FE, Nian H, et al. Constrictive bronchiolitis in soldiers returning from Iraq and Afghanistan. N Engl J Med. 2011;365(3):222–30. https://doi.org/10.1056/nejmoa1101388.

Pulmonary Fibrosis: End Game for the Pigeon Breeder

9

Dennis L. Caruana

Abbreviations

BMI	Body mass index
CT	Computed tomography
GERD	Gastroesophageal reflux disease
IgE	Immunoglobulin E
SOB	Shortness of breath

Case Presentation

A 53-year-old Caucasian woman with a history of asthma and gastroesophageal reflux disease (GERD) presented to her primary care physician with concerns of cough and worsening shortness of breath (SOB), the latter occurring in association with anterior chest pressure. She is a lifelong nonsmoker and denies regular consumption of alcohol. She remembered her asthma beginning when she was 8 years of age, but she had not required any form of treatment until adulthood. She shared with us that her asthma

D. L. Caruana (✉)
Yale University School of Medicine, New Haven, CT, USA
e-mail: dennis.caruana@yale.edu

A. M. Szema (ed.), *Unusual Diseases with Common Symptoms*,
https://doi.org/10.1007/978-3-031-40280-7_9

symptoms are mostly seasonal and were well-controlled with the help of her albuterol rescue inhaler until March. Now a month later, she had become progressively short of breath, which she and her primary care physician initially attributed to worsening asthma, and her rescue inhaler was no longer cutting it: She now had exertional breathlessness when walking on level ground for as short a distance as 100 ft. She even revealed to us that her breathlessness was sometimes occurring at rest.

The cough with which she was now presenting was nonproductive; that is, it is not producing any gunk. Now, what do I mean by *gunk*? I really mean that when she coughs, she is not coughing anything up; this includes blood (also called *hemoptysis*) and mucus, which could be gray, green, yellow, or rusty in color—and still other variations are possible. Although she was not experiencing any coughing that would awaken her at night, the cough was described as being worse first thing in the morning (i.e., upon waking) and in the midafternoon. She reported that she had an upper respiratory infection in January of this same year (2014) but denied having been sick since then. At the initial encounter, her primary care physician believed this presentation—that of increasing shortness of breath or SOB (or *dyspnea*) and nonproductive cough not associated with any signs that would suggest an infectious etiology—to be most compatible with, and after an unsuccessful trial of prednisolone, she was placed on Symbicort®, an inhaler for persistent asthma that contains two medications: *budesonide* and *formoterol*. This combination is intended to provide lasting relief of SOB by budesonide-mediated reduction in airway inflammation and by formoterol-mediated dilation of airways—which is the same mechanism as her albuterol rescue inhaler mentioned earlier with the caveat that formoterol sustains airway dilation for longer than albuterol. Note, combining an inhaler whose effects are long-lasting, such as Symbicort®, with an albuterol rescue inhaler allows for reduced frequency of asthma attacks (i.e., asthma attacks will not occur as often) and the ability to manage symptoms of asthma attacks (e.g., SOB and wheezing) when they strike, respectively. Getting back to our patient, addition of the Symbicort® inhaler provided only modest improvement. What else could be going on?

It is worth noting that our patient also complained of chest heaviness and had a history of GERD, a disease characterized by reflux of stomach acid into the esophagus. The esophagus was not designed to receive acid from the stomach. Consider for a minute that one of the most effective strategies employed by our innate immune system—the branch of our immune system that is general in its response to pathogens and lacks immunologic memory (i.e., knowledge about which cell surface proteins were most ideally suited to defend against specific pathogens that our immune system has seen before from previous skirmishes)—is to kill microbes that invade our upper respiratory tract and oral cavity by having us swallow such invaders, ultimately exposing them to the caustic environment of the stomach. We can then imagine that this same caustic environment spilling over into the esophagus could reasonably be predicted to cause irritation of the esophageal lining.

Despite her current regimen of omeprazole, a medication that reduces acid production in the stomach, she was still experiencing occasional reflux. Her primary care physician also observed that she has gained a "moderate amount of weight" secondary to dietary indiscretion and documented her body mass index (BMI, a measurement of one's mass per one's height) as 37.8 kg/m^2 (obesity: BMI > 30.0) but additional information (e.g., how much weight the patient has gained) was not made available. Accumulation of fat around the abdomen (subcutaneous fat) and inside the abdominal cavity (visceral fat) increases the pressure in the abdominal cavity, which favors efflux of stomach acid into the esophagus by pressing the contents of the stomach through the gastroesophageal sphincter—a sphincter between the stomach (*gastro-*) and esophagus (*–esophageal*) that is normally contracted (closed) and relaxes (opens) only to allow passage swallowed substances into the stomach. Although unlikely to be helping with her GERD, she had endorsed regular consumption of 1 cup of coffee per day for which we encouraged discontinuation—coffee consumption reduces the tone of the lower esophageal sphincter, thereby facilitating reflux of stomach acid into the esophagus.

Classically, however, acid reflux of GERD causes burning pain in the epigastric region (i.e., the area of the abdomen that overlies the stomach) and is associated with belching, whereby one can often taste refluxing acid. Since her discomfort is atypical for GERD. When asked to describe the quality of the chest pain again, she described it as a *chest tightness*, which contrasts with its initial character—*anterior chest pressure*. Anterior chest pressure, particularly when the patient clarifies the pressure to be *crushing* and/or *substernal* (i.e., localizing to directly beneath the sternum) and describes the pain as *radiating* (traveling), especially to either the left arm or neck, would have increased our concern for angina because it would be more consistent with an acute coronary syndrome. Her primary care provider also notes that there was no edema on our patient's physical exam. This, in combination with our patient's cough and SOB occurring over the span of a month and no recent history of travel involving prolonged immobilization or surgery reduced our concern for deep venous thrombosis leading to pulmonary embolism (i.e., formation of a blood clot in the deep veins of the leg and fragmentation of the clot that lodges in the pulmonary vasculature, respectively).

Regarding our patient's social history, she has been employed for several years as a veterinary technician and reports tending to mice, rats, turkeys, rabbits, guinea pigs, dogs, cats, chimpanzees, and sheep intended for use in basic science research. She also had many pets at home—including dogs, cats, and several birds—stating that she had always been around animals at work and at home and denying that there had been any recent changes in either the number or type of animals to which she is exposed in either setting. Despite this, her primary care physician persisted; she revealed that the birds she was keeping in the home were pigeons. Once privy to this information, he began her on 60 mg prednisone daily to reduce inflammation presumably associated with… *drumroll please*……**pigeon breeder lung disease!** Pigeon breeder lung disease is a form of **hypersensitivity pneumonitis** (a fancy way of saying inflammation of the lungs [*pneumonitis*] due to immunologic oversensitivity [*hypersensitivity*] to an environmental allergen or irritant)—and he strongly recommended that she remove the pigeons from her home at once.

Remarkably, removal of the birds in the home led to complete resolution of our patient's cough and SOB within a few days. It is also worth noting that she continued to work as a veterinary technician which was felt may put her at risk of pneumonitis—and potential recurrence of respiratory symptoms—secondary to exposure to other environmental allergens, such as those present in dust in the enclosures of other birds. We must at this time note that pigeon breeder lung disease is not limited in etiological scope to allergens present in the dust of pigeon droppings, as is classically the case; rather, people can have immunological hypersensitivity to a multitude of airborne avian antigens derived from a myriad of bird species not limited to pigeons. It is for this reason that pigeon breeder lung disease actually encompasses a more general inhalational lung injury that can occur secondary to immunologic hypersensitivity and is also known as *bird fancier's lung*.

When she was referred to our office for allergy and pulmonary function testing—roughly a year after she first presented to her primary care physician—we began with testing for hypersensitivity to a panel of common environmental allergens as well as with spirometry, the test of choice to assess for defects in ventilation associated with the large airways. Our patient tested positive for guinea pig, rat, rabbit, dust mite, weeds, and grass. Notably, she tested negative to chicken, duck, and goose feather allergens. At this time, her spirometry revealed a mild obstructive ventilatory defect with only mild improvement after bronchodilator (Fig. 9.1, Table 9.1); in plain English, this means she has a difficulty evacuating air from her lungs and only had slight improvement in doing so with the help of a medication that increases the caliber of her respiratory passages. This is compatible with her history of asthma, one of the most common forms of obstructive lung disease—the other being chronic obstructive lung disease. Her asthma symptoms (e.g., wheezing, SOB) were well controlled for another year. What happened?

On review of systems, her cough was noted to be chronic and dry. She also reports becoming increasingly SOB and experiencing occasional chest pain with breathing but cannot pinpoint exactly when this all began. She states it has been gradual in

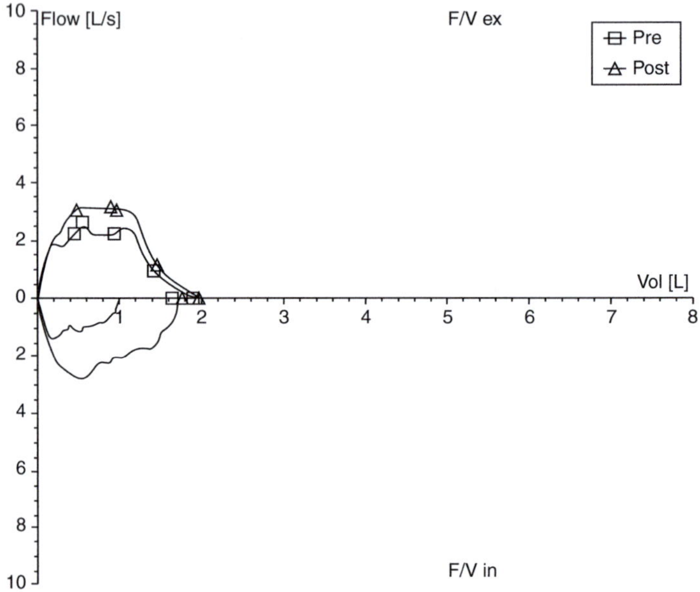

Fig. 9.1 Spirometry *Pre* pre-bronchodilator, *Post* post-bronchodilator, *Vol* volume

Table 9.1 Spirometry, September 2015

	Predicted	Pre-BD	%(Pre/ Predicted)	Post-BD	%(Post/ Predicted)	%(Post/ Pre)
FVC	3.10	1.89	61	1.96	63	103
FEV1	2.43	1.64	67	1.77	73	108
FEV1%M	79	86	109	91	114	105
FEF25-75%	3.10	2.02	65	2.68	87	132

BD bronchodilator, *FVC* forced vital capacity, *FEV1* forced expiratory volume in 1 s, *FEV1%M* percent of the FVC accounted for by the forced expiratory volume in 1 s (i.e., 100% * FEV1/FVC), *FEF25-75%* forced expiratory flow over the middle half of the FVC onset, over several months. What could be going on? What a perfect opportunity to *talk to the patient* and to *gather more information*!

Her exposures now included animal exposures, potentially to airborne fecal matter, and dust from ongoing construction in her workplace that allegedly lingers as a result of poor ventilation. Aside from her pets—she at this point still had six dogs and six cats—her home exposures were significant for asbestos and the dust of industrial metals. Our patient had been living in housing located in Calverton, New York that had been formerly subsidized and reserved for employees of the now-defunct Grumman Aerospace Corporation, whose claim to fame arguably was their engineering of the Apollo Lunar Module and production of military as well as civilian aircraft. When she first moved into her housing unit, she noted that there was a film of metallic dust on the walls and ceilings, presumably from industrial workers bringing the dust home on their clothing.

After learning all of this, we ordered a chest X-ray and computed tomography (CT) scan of her chest to *take a look under the hood*, as is commonly the expression. At this time, nearly 2 years after worsening of her asthma symptoms were thought to have arisen secondary to hypersensitivity pneumonitis, her chest X-ray demonstrated infiltration of the lingula and lower lobe of the left lung with diffuse bronchial wall thickening. The radiologist read this as suggestive of pneumonia, yet our patient denied having a recent history of systemic signs of illness, such as fatigue, fever, and chills; denied having any sick contacts; and was constitutionally well-appearing on physical exam. Although gradual onset certainly does not exclude the possibility of an acute pneumonia superimposed on insidious development of an underlying lung disease, the time course of our patient's symptoms does not fit well with the narrative of pneumonia. Her chest CT scan showed interstitial lung disease with some scattered ground-glass opacities more pronounced in the lingula and lower lobe of the left lung as compared to that of the right lung. The radiologist was clear as day in his impression: "I suspect underlying pneumonitis."

Diagnosis

Our patient underwent wedge biopsy of the upper lobe of the left lung (aka left upper lobe [LUL]) that was sent for histopathological interpretation and showed the following: chronic bronchiolitis (i.e., longstanding inflammation involving the small airways), peribronchial metaplasia (i.e., reversible change in the type of cells from that which normally predominates in a given tissue to another cell type that is part of the tissue adaptive response to chronic irritation but is usually detrimental; a classic example of metaplasia is the loss of ciliated lining of the respiratory tree in smokers in favor of a more rugged lining that predisposes to infection secondary to reduced microbial clearance), mucostasis (trapping of mucus in the small airways—literally, the stasis of mucous), and bronchiectasis (permanent dilatation [increase in caliber] of the small airways) as well as patchy airway centered and peripheral interstitial fibrosis (i.e., patchy scarring of the lung). Given our patient's history of asbestos exposure, we also considered it necessary to rule out asbestosis—a restrictive pattern of lung disease that usually manifests as a fibrotic interstitial lung disease with a usual interstitial pneumonia-like pattern. This pattern was not observed in the biopsy and no asbestos bodies (filament-like structures that suggest one has inhaled asbestos) were noted.

What does all of this mean? And, most importantly, what does all of this mean for our patient? First, a wedge biopsy is a method of sampling the lung whereby a surgeon removes a *wedge* of tissue for review by a pathologist. The pathologist interpreted the sample they received as compatible with the above findings, which represent small airway disease and irreversible scarring of the lung. In a manner similar to how statistics draws conclusions from a sample for which collection of data is logistically feasible and conclusions made on the basis of these data extrapolated to a population, if we consider the pathological findings which correspond to the wedge biopsy sample to be representative of a much larger chronic inflammatory process involving lungs—as suggested by our previous chest CT scan showing diffuse ground-glass opacities—we reasonably infer that our patient has a progressive and irreversible restrictive pattern of lung disease that most likely arose from hypersensitivity pneumonitis leading to eventual interstitial pulmonary fibrosis.

Treatment

The only definitive treatment option for our patient is lung transplant. Unfortunately, her BMI of 37.7 kg/m^2 caused her to be ineligible for lung transplantation at the time of referral. Frankly, her obesity (BMI > 30.0 kg/m^2) is such that the risk of harm associated with transplantation outweigh the potential benefits thereof. We then counseled the patient regarding nutrition, regular exercise, and weight reduction in order to reach a goal BMI of 34, at which point we plan to reconsider lung transplantation.

In the meantime, since our patient plans to continue working as a veterinary technician, we have advocated on her behalf to ensure she equipped with a respirator such that she may minimize her inhalation of allergens which would otherwise exacerbate her hypersensitivity pneumonitis; this will, in theory, slow progression of her pulmonary fibrosis by limiting inflammation associated with allergic hypersensitivity.

Key Points

1. Interstitial pulmonary fibrosis is an insidious restrictive process whose development occurs over the span of years and may present only as progressively worsening SOB ± cough.
2. Although hypersensitivity pneumonitis is treatable by avoiding exposure to inciting environmental allergens to which the patient has hypersensitivity, interstitial pulmonary fibrosis may arise as a complication of hypersensitivity pneumonitis.
3. The only definitive cure that is currently available for interstitial pulmonary fibrosis is lung transplantation.

Multiple Choice Questions

1. Which of the following is *NOT* a form of restrictive lung disease?
    (a) Idiopathic pulmonary fibrosis.
    (b) **Asthma.**
    (c) Hypersensitivity pneumonitis.
    (d) Pneumoconiosis.

2. Which test is preferred for the assessment of large airway disease?
    (a) Impulse oscillometry.
    (b) **Spirometry.**
    (c) Exhaled nitric oxide.
    (d) Methacholine challenge.

Answers

1. (b)
2. (b)

Handslapping the Table May Be a Symptom Worth Investigating

10

Mosammat Perveen

Vignette

Ms. Kleinfaser walked into the office in hopes of finally resolving the constant pain she had been enduring for the last couple of years. At 55 years old, Ms. K had been suffering from excruciating pain in her hands and legs. K had already been through a lot in her life and these symptoms exacerbated her stress along with her pre-existing medical conditions. Due to an unfortunate car accident, she broke her wrist in 2013 and was diagnosed with shingles the following year. Ever since then, she has been riddled with pain and burning sensations in her hands and feet. Her hands would crawl and itch almost as if there were thousands of needles in her skin. At times she would get frustrated to the degree where she would deliberately bang her hands on tables and walls to relieve the pain. However, this was not all of K's worries. She had noticed that lately she had been feeling more tired than usual. K was confused; she woke up and went to bed everyday feeling fatigued and tired. Even eating was becoming uncomfortable for her. Her stomach would churn and jolt every time she ate. She knew she

M. Perveen (✉)
Department of Epidemiology, School of Public Health,
SUNY Downstate Health Sciences University, Brooklyn, NY, USA

Kentucky College of Osteopathic Medicine, Pikeville, KY, USA

A. M. Szema (ed.), *Unusual Diseases with Common Symptoms*,
https://doi.org/10.1007/978-3-031-40280-7_10

had underlying thyroid problems, but attributing her recent symptoms of burning pain, fatigues, and discomfort to her previously managed health problems did not make sense. Her deteriorating health took a heavy toll on her as she started to quickly gain weight. K ultimately slipped into depression. Her self- esteem had dropped and she no longer went outside, partake in hobbies, or even socialized with her friends and family. The pain and stress became too much for K to even get out of bed. She gained weight. Her constant fatigue kept her bound to her home. K wanted to relieve herself of this pain and most importantly get her old life back. After her previous doctors failed to come to a diagnosis, she was referred to allergist Dr. Szema, in hopes that a diagnosis and treatment can soon be reached.

When Ms. K presented herself to Dr. Szema, she was experiencing lots of fatigue in the mornings and nights along with paresthesia, a painful burning sensation in her hands and legs. She slapped the table, she says, "in order to get rid of the pain." Repeatedly she kept on slapping the table with her hands. Over the past couple of months, she had gained a lot of weight and had an accompanying rash. Her medical history revealed idiopathic urticaria, Hashimoto's thyroiditis, asthma, and hypertension. Hashimoto's disease is an autoimmune disease that causes inflammation of the thyroid gland. She developed two thyroid nodules, weight gain, and very dry skin. K already had urticaria, a chronic form of hives that resulted in occasional itchy flare ups. She was already under omalizumab(anti-IgE) administration and observation to treat her urticaria, a series of injections over the span of 2–4 weeks. Nevertheless, KN's skin was under a lot of stress and coupled with her weight gain, her mental health suffered along with her physical. The feeling of pins and needles up and down her arms discouraged her from going outside, hence she stayed home for weeks at a time. Blood tests revealed that Vitamin D levels had dropped due to lack of sunlight. After reviewing her previous blood tests and cytology reports, we concluded that K had her asthma, hypertension, and Hashimoto's thyroiditis under control. Her thyroid nodules were benign and her rash was controlled with antihistamines. Yet her paresthesia still subjected her to immense pain and burning sensations.

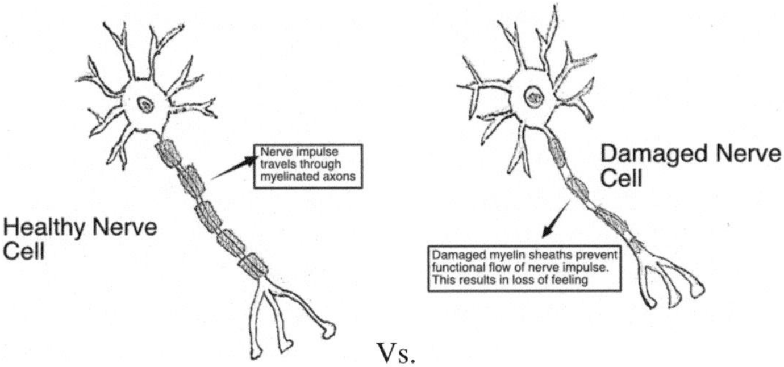

While going through a plethora of tests and medical records, Dr. Szema came across her nerve fiber density test. The biopsy revealed that KN had reduced intra-epidermal nerve fiber density, a strong marker for diagnosing small fiber neuropathy (SFN). SFN is characterized by damage to the peripheral nerves thus causing neuropathic pain [1]. This pain presented itself as paresthesia in KN's hands and feet often subjecting her to sensations of burning pain or pins and needles. After understanding the cause behind KN's symptoms, Dr. Szema proceeded to treat her with an intravenous immunoglobulin treatment (IVIG). Within a week KN started to feel better, her pain had decreased and she no longer felt fatigued when waking up or going to bed. The relief had improved her mental health and she was ready to jump back into her life. Her intravenous therapy took 3 days and she had to come back and repeat the treatment every 3 weeks. However, to KN, this was a very small price to pay. The anguish and burden of not knowing the causes behind her pain and how to stop it was much worse than a couple more trips to the doctor's office. KN started socializing again and spending time with her family and friends. With supplements and ample time outside in the sunlight, her Vitamin D levels were also starting to increase. After a long while she finally started to feel some relief. After 2 weeks she felt better than she ever had during these past years. After years of dead ends and lack of explanations, KN finally managed to get a piece of her life back.

Background/Salient Features of the Case

Small Fiber Neuropathy

Small fiber neuropathy occurs as a result of peripheral nerve damage involved in somatic and autonomic functions. Symptoms include burning pains or sensations similar to pins and needles, gastric problems, dry mouth and eyes, abnormal sweating, fatigue, and syncope [2]. Damage to the myelinated A fiber nerves and unmyelinated C fiber nerves can result in the loss of small nerve fibers or display abnormal nerve structure in the upper and lower extremities [2]. These fibers are primarily mechano thermal and polymodal nocireceptors. Hence they warp the sensations of pain and tingling both with and without contact to external stimuli [2]. The disease often creates diagnostic confusion as symptoms and markers are not entirely clear. Patients are almost always neurologically asymptomatic with the exception of sensory and receptive symptoms evident of autonomic dysfunction. These can range from dry mouth, dry eyes, gastric problems, abnormal sweating, and dry cracked skin in certain affected areas. The most common cause of developing small fiber neuropathy is diabetes mellitus; however, other physiological factors play a significant role in increasing the risk factor such as hyperlipidemia, abnormal glycolytic metabolism, and autoimmune conditions such as Hashimoto's thyroiditis [3]. Also included are genetic diseases such as Fabry's disease and other non-hereditary diseases such as HIV, celiac disease, and hepatitis. Although the pathology behind small fiber neuropathy is not fully understood, the commonality among most patients is the intense instances of pain associated with other sensory symptoms.

Diagnosis

KN had extensive workup done with multiple doctors in order to pinpoint the cause of her pain. Numerous blood tests, cytology reports, and sonograms were conducted to rule out the cause as her Hashimoto's thyroiditis and low vitamin D levels. It was not until a skin biopsy with Dr. Klein did the true diagnosis come forth. The skin biopsy was taken for a nerve fiber density test (epidermal nerve fiber density or ENFD)) and KN's test showed that she had reduced intraepithelial nerve fiber density. KN had a lower than normal number of epidermal nerve fibers than the average person. This is a hallmark of small fiber neuropathy and the ENFD test itself is a common practice to assist in diagnosing [4]. The biopsies are subjected to high powered magnification to assess the dilation of the small nerve fibers in the skin. This allows health professionals to visualize and calculate how many fibers are present per surface area. A low density will indicate that there are reduced fibers or some have been damaged [5]. Like KN, patients often feel a version of sensory disruption. Loss of density or damaged nerve fibers are what causes the abnormal peripheral functions, such as KN's burning sensations in her hands and feet.

Treatment

Treating SFN includes addressing the underlying cause for its development. Some causes include diabetes mellitus or patients who are pre-diabetic and unable to manage their blood sugar. In these cases, managing diet and using insulin to control blood sugar usually results in a decrease of symptoms. However, there are many cases of SFN in which patients do not suffer diabetes. KN did not have diabetes nor did her blood tests show any indication of prediabetes. Some cases of small fiber neuropathy are considered "idiopathic" or in simpler terms, cause is unknown. Another viable proposal is an autoimmune response resulting in SFN. Autoimmune responses involve the patient's immune system targeting its own tissues causing damage to the peripheral fibers. Since KN was not diabetic and did not show any indication of prediabetes, the cause of her SFN was narrowed to idiopathic and autoimmune causes. To treat this type of SFN, an immunomodulation therapy proves to be highly efficacious. KN was treated through intravenous delivery of immunoglobulin (IVIG). IVIG is an antibody mostly responsible for humoral and pathogenic immunity. KN was administered IVIG-in high doses to decrease the potential immune response causing her SFN and act as an anti-inflammatory agent. KN immediately felt better and her pain had reduced substantially. After almost 2 weeks, KN told Dr. Szema she had not felt this good in months. As of now, treatment for SFN is still being studied since not much is known about the underlying causes. However, IVIG is proving to be a potential therapeutic treatment to alleviate pain in SFN patients. We also now administer immunoglobulin G subcutaneously, so we do not need to use an intravenous line.

Fun Facts

1. Almost 40 million people in the USA are currently diagnosed with either small or large fiber neuropathy.
2. Chances of developing small fiber neuropathy are approximately 52 per 100,000 people.
3. Prevalence of small fiber neuropathy is higher in men and the elderly.

Multiple Choice Questions

1. Which of the following is a viable cause for developing small fiber neuropathy?
   (a) Diabetes.
   (b) Asthma.
   (c) Kidney stones.
   (d) Poor diet.

2. Which of the following is the appropriate procedure to diagnose small fiber neuropathy?
   (a) Blood test.
   (b) MRI scan.
   (c) Nerve fiber density test.
   (d) Colonoscopy.

3. Which of the following conditions is an autoimmune disease?
   (a) Alzheimer's.
   (b) Osteoporosis.
   (c) Arthritis.
   (d) Hashimoto's thyroiditis.

Answers

1. (a)
2. (c)
3. (d)

References

1. de Greef BTA, et al. Intravenous immunoglobulin therapy for small fiber neuropathy: study protocol for a randomized controlled trial. Trials. 2016;17(1):330. https://doi.org/10.1186/s13063-016-1450-x.
2. Hovaguimian A, Gibbons CH. Diagnosis and treatment of pain in small-fiber neuropathy. Curr Pain Headache Rep. 2011;15(3):193–200. https://doi.org/10.1007/s11916-011-0181-7.
3. Magazine, IG Living. "IVIG for Small Fiber Neuropathy." Nufactor, www.nufactor.com/patients/resources/ig/articles/ar-ivig-for-small-fiber-neuropathy.html.
4. McArthur JC, Stocks EA, Hauer P, Cornblath DR, Griffin JW. Epidermal nerve fiber density: normative reference range and diagnostic efficiency. Arch Neurol. 1998;55(12):1513–20. https://doi.org/10.1001/archneur.55.12.1513.
5. Bako Diagnostics. Epidermal nerve fiber density testing : an overview. Alpharetta, GA: Bako Diagnostics; 2017. bakodx.com/epidermal-nerve-fiber-density-testing-an-overview/.

The Nasal Trifecta

11

Mohammad Shigri, Anthony M. Szema, and Robert A. Promisloff

Vignette

In late March of 2018, a male in his late 40s, Paul Lups, came into the office with a chief complaint of a swollen lip from taking aspirin. Mr. Lups informed us at the visit that he is an electrical contractor. In his profession, he is exposed to a variety of job sites ranging from warehouses to dirt pits to old vents and sheds. This means that he is bombarded with many potential allergens such as dust mite antigens and mold in dirt. Mr. Lups had been hard at work, during the day, helping lay down the foundation for electrical lines in the ground outside in the grueling weather. Mr. Lups was exhibiting symptoms of nasal polyps earlier in the year and decided to see an Ear Nose Throat (ENT) surgeon so that he may immediately seek resolution. The ENT surgeon would remove the polyps diagnosed earlier that month and refer him to Dr. Szema to help prevent something similar from occurring again.

M. Shigri (✉)
New York College of Podiatric Medicine, New York, NY, USA

A. M. Szema
Donald and Barbara Zucker School of Medicine at Hofstra/Northwell, Hempstead, NY, USA

R. A. Promisloff
Medicine, Drexel University College of Medicine, Bryn Mawr, PA, USA

A. M. Szema (ed.), *Unusual Diseases with Common Symptoms*,
https://doi.org/10.1007/978-3-031-40280-7_11

Mr. Lups has a past medical history of asthma and bronchitis and the polypectomy procedure completed earlier that month. During the time of his visit he had not taken any antihistamines in the last 7 days. He denied food allergies but was allergic to cockroach, mold, dust, penicillin, and even aspirin. He was so sensitive to aspirin that he would break out in hives with swelling lips and a scratchy throat. This adverse reaction to aspirin is commonly referred to as aspirin exacerbated respiratory disease (AERD) or Samter's triad.

During his examination, the doctor recommended that Mr. Lups complete a series of tests to obtain a better picture of his current condition. He was first given a self-examination survey form for his asthma symptoms (Asthma Control Test or ACT), which displayed that he had well-controlled asthma. Next, he was given an exhaled nitric oxide test for his asthma. This test would measure levels of nitric oxide in his breath, which is used as an indicator for inflammation in patient airways. In the last respiratory test, Dr. Szema would do a spirometry/bronchodilator study. Mr. Lups had been able to keep his asthma under control but still had lingering nasal problems that he told Dr. Szema about. He said that he often felt congested in both his nostrils and would have to go about his day with a persistent stuffy nose. On top of that, he would feel pain in his right nostril. He recalled with sorrow how he progressively lost the ability to smell but is still thankful that he can taste food to an extent. With the data at hand, Dr. Szema would go on to conclude that the most effective method to help Mr. Lups would be aspirin desensitization.

Background/Salient Features of the Case

Aspirin Exacerbated Respiratory Disease

The first known case of adverse reaction to aspirin was documented in 1902 and the disease was studied in 1922 by Ferdinand Widal. Widal and colleagues were able to demonstrate the effects of the disease with help from volunteers ingesting aspirin and

NSAIDs. Although publishing his study in the French journals in 1922, the study did not get much recognition until Samter and Beers released their own study in 1968 [1]. Patients with AERD are often prescribed acetaminophen alternatives at a low dose (up to 500 mg) [2]. Although many medical therapies are available patients still often need nasal polyp surgeries. Physicians recommend aspirin desensitization shortly after the procedure to reduce future nasal polyp growth and recurrence [1].

Aspirin exacerbated respiratory disease (AERD) is a medical condition that is most classified by a triad of symptoms including, asthma, nasal polyps, and a reaction to aspirin. These core symptoms, made famous by Max Samter, is the reason that disease is also commonly known as Samter's Triad. AERD can be a tricky disease to navigate as it manifests suddenly during adulthood (although any age group can have it) and can have a wide range of symptoms from skin reactions to vomiting [1]. AERD also requires a clinical diagnosis, which means that currently there is no one test that can determine whether you have it and relies on the patient's symptoms and medical history [2].

Once the diagnosis of AERD is confirmed, one of the first treatment options are leukotriene modifiers. Leukotrienes are inflammatory compounds which are produced by leukocytes (white blood cells) which induce inflammation, leading to vascular permeability—hence a runny nose. Patients with AERD produce high levels of leukotrienes which are accentuated by consuming aspirin and non-steroidal anti-inflammatory drugs (NSAIDs), such as Ibuprofen [1]. Two FDA-approved leukotriene modifiers are: (1) Zileuton, which is a 5-lipoxygenase inhibitor that blocks leukotriene formation, and (2) Montelukast, a leukotriene B4 receptor antagonist which blocks leukotriene function. The most common medication prescribed to patients is montelukast. Zileuton is an effective alternate to montelukast but it is prescribed less frequently. This may be due, in part, to cost, more frequent dosing, and the necessity to get liver function testing. Montelukast recently received an FDA black box warning for potential cognitive/behavioral dysfunction.

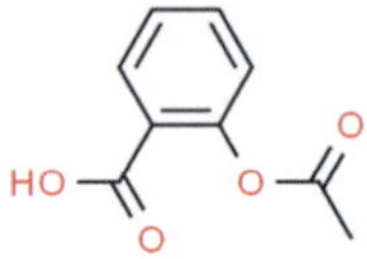

The chemical structure for aspirin

Diagnosis

The diagnosis of aspirin exacerbated respiratory disease (AERD) is derived from a patient having three main concurrent symptoms. The triad of symptoms includes asthma, nasal polyps, and sensitivity to aspirin and other NSAIDs. AERD can be extremely difficult to diagnose because of the wide variety of symptoms that may be presented as well as the fact that those symptoms develop over time. Many patients with AERD may be diagnosed later in their lives due to physicians treating symptoms independently rather than as a cause for AERD. When a patient presents with a reaction to aspirin/NSAIDs physicians may ask the patient to try an aspirin challenge to confirm it.

The aspirin challenge involves exposing the patient to aspirin and monitoring the reaction. The procedure will be done under the care of a respiratory specialist in a controlled setting with adequate equipment to ensure the patients safety. Aspirin challenge is a useful tool in identifying if a spontaneous reaction to aspirin has been developed as well as testing if other allergy symptoms are aggravated by aspirin. A past study illustrated the effectiveness of the aspirin challenge as it found that 42% of patients with asthma and other underlying nasal conditions tested positive despite not having reactions to NSAIDs in the past [1].

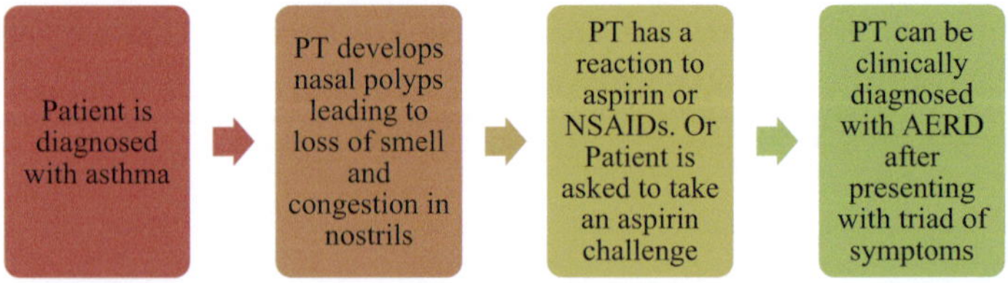

Treatment

Aspirin Exacerbated Respiratory Disease

Aspirin desensitization is commonly regarded as the most effective treatment for AERD [1]. Past clinical research has shown that 87% of patients who participated in aspirin desensitization treatment experienced improvement of their symptoms [3]. The treatment has various protocols, but the crux of it involves the patient taking doses of aspirin, starting with small doses and gradually moving up. Examples of variant protocols include: 1 day 6-h protocol, two-day protocol, and three-day protocol. When starting, it is important that the patient be in stable condition and have an FEV1, forced expiratory volume in 1 s, higher than 70% of predicted. For patient safety, the environment should also have intubation equipment, EpiPen, and a defibrillator on standby.

Mr. Lups had his aspirin desensitization treatment conducted in an ICU environment. He was treated with the 1 day 6-h protocol for aspirin desensitization. This plan starts the patient off with a 10 mg dose once and then allows 20 min for any potential symptoms to manifest. If there are no symptoms in the given time, then the patient's dose is doubled, in this case now to 20 mg, and the process is repeated until the target dose is met (162 mg). Once the patient was able to comfortably tolerate the target dose, he was sent home with 81 mg Ecotrin coated aspirin to take once a day with food.

Aspirin dose	Time of dose
Aspirin 10 mg dose	If no symptoms for 20 m, then proceed
Aspirin 20 mg dose	If no symptoms for 20 m, then proceed
Aspirin 40 mg dose	If no symptoms for 20 m, then proceed
Aspirin 81 mg dose	If no symptoms for 20 m, then proceed
Aspirin 162 mg dose	If no symptoms for 20 m, then proceed

Reference from Catherine R Weiler, MD, PhD (Mayo Clinic Rochester, AAAAI)

Discussion

On June 26, 2019 the FDA-approved interleukin 4 receptor alpha antagonist called dupilumab as treatment for adult nasal polyps [4]. A nasal polyp is a noncancerous overgrowth of the sinuses that may obstruct the ostiomeatal complex, the region of drainage. Nasal polyps are often associated with other respiratory disorders such as asthma, chronic sinusitis, and asthma. Functional endoscopic sinus surgery (FESS) was introduced in the USA in the 1980s to combat the symptoms of chronic sinusitis [5]. FESS utilizes small thin tubes with cameras called endoscopes to provide vision into the nasal pathways. There are no incisions or openings required as the surgeon will use another thin tube with surgical instruments to treat the site of disease. The surgeon will target abnormalities in the sinus to help improve drainage without damaging the cilia in the nasal mucus. The procedure may take up to 4 h and any pain can be treated with OTC medication but displays 90% improvement of symptoms [5].

This is a significant advancement in medicine as nasal polyp growth is difficult to control, often requiring surgery. Even in the event of surgical removal of nasal polyps (polypectomy) there is often regrowth that may require surgery again. Dupilumab functions by targeting the type 2 inflammatory response, inflammation that is triggered by type 2 cytokines such as IL-4 and IL-13 [4]. It blocks IL-4 function in T2 inflammation to break the positive feedback loop leading to mast cell histamine production. Type 2 is a common response in patients with asthma and nasal polyps.

Dupilumab is given as a subcutaneous injection and is recommended as a 300 mg dose every other week for nasal polyposis [4]. Dupilumab was used in a 52-week clinical study with 722 adult subjects that had chronic rhinosinusitis with nasal polyposis (CRSwNP) [4]. Patients have seen significant progress in lung function in as little as 2 weeks. Dupilumab also has some side effects including gastritis, insomnia, toothache, conjunctivitis, and arthralgia. From the clinical study referenced previously, the occurrence of adverse reactions was between 1% and 3% with injection site side effects being the outlier at 6% [4].

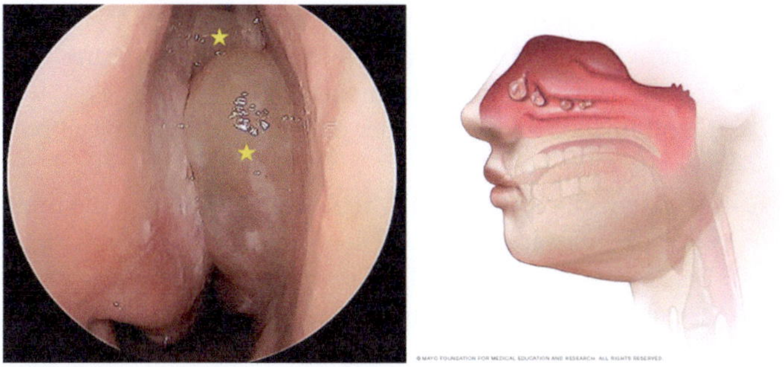

Dr. Szema will obtain a clinical image of the nasal polyp

Fun Facts

1. 20% of patients with AERD go undiagnosed, while the most common age of an AERD diagnosis is 34 [1].
2. 97% of patients with AERD report decreased or absent sense of smell, which can heavily affect the patient's romantic endeavors, as body odors subconsciously guide picking a partner [6, 7].
3. Although named aspirin exacerbated respiratory disease, patients with AERD have persistent and chronic problems, despite avoiding aspirin and other NSAIDs.

Multiple Choice Questions

1. Which following medication is the BEST treatment for nasal polyposis AERD patients?
   (a) Aspirin desensitization.
   (b) Oxymetazoline (Afrin®).
   (c) Leukotriene modifiers.
   (d) Corticosteroids.
   (e) Dupilumab.

2. What is an existing condition that may lead to the development of nasal polyposis?
   (a) Upper respiratory infection.
   (b) Emphysema.
   (c) Pleural effusion.
   (d) Cystic fibrosis.

3. Which of the following statements is true?
   (a) Improvement of FEV1 after use of albuterol is the sole requirement of asthma diagnosis.
   (b) Dupilumab is an FDA-approved medication for nasal polyposis.
   (c) A patient cannot have an AERD flare up without the presence of an aspirin or NSAID.
   (d) One of the dangers of FESS is that the patient may retain permanent damage in nasal mucosa cilia.

Answers

1. (e)

   Oxymetazoline (b) is an alpha-1 and alpha-2 adrenergic receptor agonist which is prescribed for treating nasal congestion not AERD. Corticosteroids (d) is the incorrect choice because they are commonly prescribed to relieve asthma/allergy symptoms. Corticosteroids are flexible and can be prescribed as inhalers, injections, and nasal sprays. Leukotriene modifiers (c) are also commonly prescribed as they target leukotrienes, inflammatory substances contributing to AERD symptoms, and aim to reduce inflammation in patients but are not curative. Aspirin desensitization (d) is sometimes effective treatment for combatting AERD and is shown to have significant improvement in patients' symptoms. However, the correct answer for medical treatment of nasal polyposis is dupilumab (e), which is an anti IL-4, anti IL-13 blocker that is FDA-approved for nasal polyposis.
2. (d)

   An upper respiratory infection (a) (URI) is a viral infection that can cause cough, sore throat, and nasal discharge. URIs are more frequent during fall and winter but are usually resolved within a few days, they do not cause formation of nasal polyps. Emphysema (b) is the incorrect choice as it is chronic obstructive pulmonary disease (COPD) that damages the alveoli in the lung. Emphysema is commonly acquired from smoking but can also be genetic. Pleural effusion (c) is when an excess of fluid is collected in the pleural cavity, which is the space between the lung and chest. This answer choice is incorrect as pleural effusion can occur from pneumonia or heart complications and does not commonly lead to complications in the upper respiratory tract. Cystic fibrosis (d) is the correct answer choice because chronic sinusitis commonly develops associated with CF which leads to the formation of nasal polyps.

3. (b)

   (a) is incorrect because albuterol is used as medicine for patients with asthma, so an improvement in FEV1 indicates the patient likely has asthma. (b) is the correct answer choice as patients with AERD often have high levels of leukotrienes and are prescribed dupilumab for nasal polyposis. As mentioned previously in the chapter, it is a common misconception that the presence of aspirin is required for an AERD reaction. This makes (c) incorrect. (d) is incorrect because one of the major benefits of FESS is that it preserves as many nasal cilia as possible. Studies have also shown an increase in cilia density in following weeks of the FESS.

References

1. https://www.samterssociety.org.
2. https://www.aaaai.org/conditions-and-treatments/library/asthma-library/aspirin-exacerbated-respiratory-disease.
3. Berges-Gimeno MP, Simon RA, Stevenson DD. Long-term treatment with aspirin desensitization in asthmatic patients with aspirin-exacerbated respiratory disease. J Allergy Clin Immunol. 2003;111(1):180–6. https://doi.org/10.1067/mai.2003.7.
4. https://www.accessdata.fda.gov/drugsatfda_docs/label/2018/761055s007lbl.pdf.
5. https://www.aafp.org/afp/1998/0901/p707.html.
6. Wedekind C, Seebeck T, Bettens F, Paepke AJ. MHC-dependent mate preferences in humans. Proc Biol Sci. 1995;260(1359):245–9. https://doi.org/10.1098/rspb.1995.0087.
7. Havlicek J, et al. Women's preference for dominant male odor: effects of menstrual cycle and relationship status. Biol Lett. 2005;1(3):256–9. https://doi.org/10.1098/rsbl.2005.0332.

It's Not the Dentures: It's Your Immune System Ms. Jones!

12

Anna Veintimilla

A. Veintimilla (✉)
Renaissance School of Medicine at Stony Brook University, Stony Brook, NY, USA
e-mail: anna.veintimilla@stonybrookmedicine.edu

A. M. Szema (ed.), *Unusual Diseases with Common Symptoms*,
https://doi.org/10.1007/978-3-031-40280-7_12

CVID

Vignette

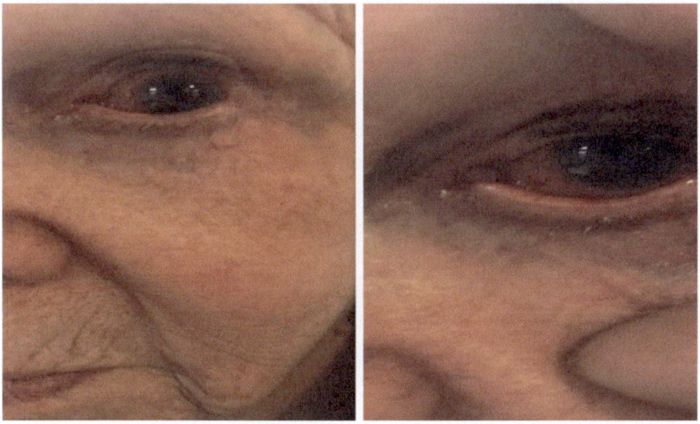

Missing teeth, bloody gums, a sore mouth—this was Ms. Jones's reality! With the condition of her mouth slowly worsening, Ms. Jones sought out help from the dean of one of the two most prestigious dental schools in the largest city in Beantown, so she could restore her once radiant smile. She believed a beautiful set of expensive acrylic dentures would be the answer to her prayers. Unfortunately, what was supposed to be a happy ending turned into a living nightmare. Ms. Jones's new smile was quickly ruined by a painful infection. Little did she know, this infection would be the first of many.

In 2013, Ms. Jones, now a 93-year-old woman, walked through the doors of Dr. Szema's office after suffering from multiple mouth infections, ultimately leading to multiple broken teeth and dentures. Furthermore, Ms. Jones also had a long history of bacterial bloodstream infections. For over a decade, she had a slowly growing lymphoma which remained untreated. She once was diagnosed with Babesiosis, a tick-borne infection, and recalled multiple episodes of bacterial pneumonia.

Members of Ms. Jones's family also suffered from certain health complications. The family has had a long history of cancer within its blood line. Her brother had lung cancer and her child had an unspecified cancer. The daughter was diagnosed with common variable immune deficiency (CVID), and just like her mother, she suffered through multiple episodes of pneumonia. In regard to her parents, Ms. Jones's father has diabetes and her mother noted a history of hypercholesterolemia.

Why is it that Ms. Jones kept experiencing these recurring infections? Was age the determining factor here? As you hit a certain age, do you become more susceptible to infections? It is true that as one ages they will develop more comorbid conditions, such as arthritis. Along with comorbid conditions, a person's immune activity will also tend to decrease with age. This cocktail of comorbidities along with a weaker immune system can leave a person more vulnerable to infections. However, Ms. Jones's recurring battle with mouth infections was not necessarily caused solely by her age. Like mother, like daughter, the main source behind Ms. Jones's recurring visits to the doctor's office can be attributed to hypogammaglobulinemia, or more commonly called, common variable immune deficiency (CVID). Patients with this condition are unable to make enough antibodies called immunoglobulins such as IgA, IgM, IgE, and IgD [1]. Normally, these antibodies are responsible for neutralizing a pathogen which may enter the body making it easier for that said pathogen to be destroyed (Fig. 12.1). This deficiency in immunoglobulins leaves a patient more vulnerable to a pathogenic attack causing recurring infections (Fig. 12.2). Patients with CVID often remain undiagnosed. The conventional way to make a diagnosis is via a blood test which measures total IgG immunoglobulin serum levels. These IgG serum levels are typically less than or equal to 50% the lower limit of normal in association with two episodes of pneumonia.

Her IgG levels were seemingly "normal" measuring within the reference range of 635–1741 mg/dL at 741.5 mg/dL (Table 12.1).

Her IgA level was low: 14 mg/dL (range of 70–400 mg/dL) (Table 12.1).

To the uninitiated, the answer to her complications seems simple in nature. Treat Ms. Jones with IgA to increase her level. The answer, however, is a bit more complex. It is fraught with peril to treat a patient who has CVID with IgA because if their body produces even the slightest amount of allergic antibody called IgE, it could potentially induce anaphylaxis. Even though her IgG level was within normal range, this concentration was not sufficient to fight her sundry recurrent infections. To bring her IgG serum level to an appropriate

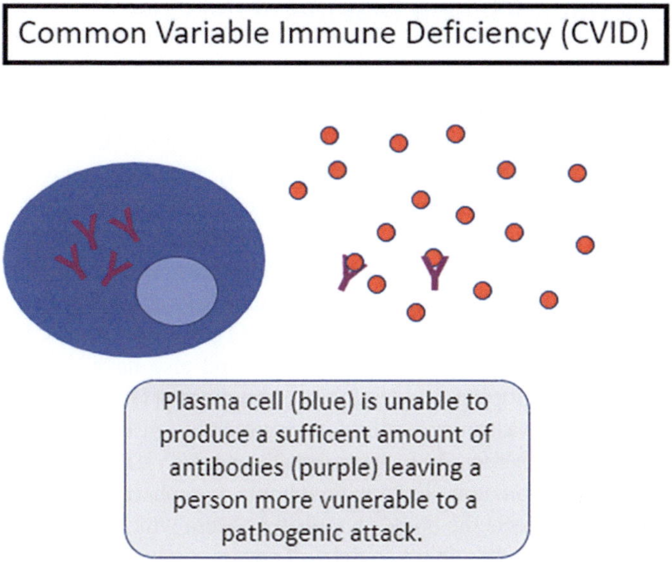

Fig. 12.1 CVID

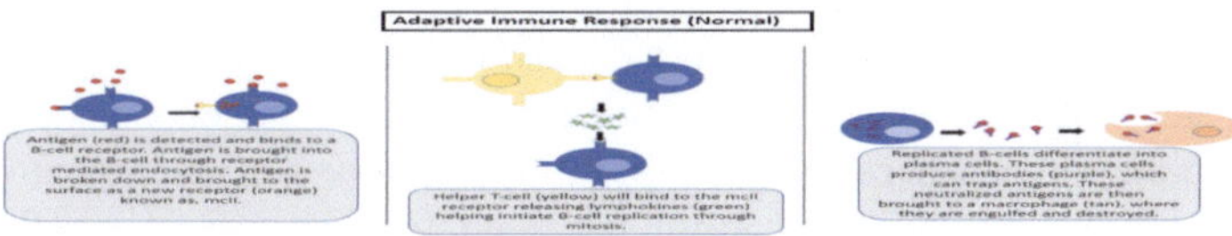

Fig. 12.2 Adaptive immune response

Table 12.1 Ig serum levels of Ms. Jones

Ig type	Serum levels mg/dL	Reference range mg/dL	Status
IgG	741.5	645–1741	Normal
IgA	14.1	66–433	Low
IgM	57.6	45–281	Normal

level immunoglobulin therapy, known as immunoglobulin infusion 10% subcutaneous (human with recombinant human hyaluronidase), was used. As far as infections go, many patients who suffer from CVID are usually treated with some form of antibiotic that will attack the bacterial infection that is present [2].

Background/Salient Features of the Case

What other symptoms come along with CVID? Besides experiencing recurring infections, a patient with CVID could develop a second autoimmune disorder due to the impairment of the immune system. This impairment results in the body attacking its own tissues and organs. This means that a patient suffering from CVID could develop disorders such as immune thrombocytopenia—an abnormal bleeding disorder caused by a decrease in blood clotting cells, or even autoimmune hemolytic anemia—a condition caused by the destruction of of red blood cells [1]. Patients with CVID could even develop rheumatoid arthritis and also have a greater chance of developing non-Hodgkin's lymphoma [1].

Ms. Jones was no exception to the possibility of developing an autoimmune disorder. As is the case with many CVID patients, Ms. Jones suffers from non-Hodgkin lymphoma. Lymphoma is a class of cancers that affects the body's lymphatic system which plays a crucial role in the body's ability to fight off any invading pathogen that is potentially harmful. Lymphoma can start anywhere in the body where there is lymph tissue such as the lymph nodes, spleen, and bone marrow [4]. There are two possible non-Hodgkin's lymphomas that a patient can have, and both types depend on the type of lymphocyte that is affected. Patients with B-cell lymphoma experience abnormalities in their B-cells [4]—a type of lymphocyte that creates antibodies that attach to the harmful pathogen marking it so other cells know to destroy the invader. Patients with this type of lymphoma may be unable to produce an adequate amount of antibodies, which results in recurring infections. Another type of non-Hodgkin lymphoma is T-cell lymphoma [4] whereby the patient experiences abnormalities within their T lymphocytes. These lymphocytes can have multiple roles, such as directly attacking the invading pathogen or helping boost/slow down the activity of other immune cells. Ms. Jones suffers from B-cell lymphoma which has been slowly growing and is carefully monitored. Considering the fact that lymphocytes (white blood cells) are an essential component of the immune system, it makes sense that a patient, such as Ms. Jones, who has CVID, would be more susceptible to developing cancers that involve components of the immune system, such as non-Hodgkin's lymphoma [4].

Diagnosis

After discovering low IgG serum levels through a blood test, Ms. Jones was diagnosed with common variable immune deficiency (CVID), also called hypogammaglobulinemia. This autoimmune disorder affects the body's immune response by affecting the production of antibodies which are crucial in helping fight off any pathogenic attack. As a result of this weak immune response, Ms. Jones is highly susceptible to recurring infections within the lungs, ears, sinuses, etc. Due to her condition, Ms. Jones is much more likely to develop other autoimmune disorders and certain cancers. Ms. Jones was also diagnosed with non-Hodgkin's lymphoma which could be linked with her CVID condition. To be more specific, Ms. Jones suffers from B-cell lymphoma where her B-cells are unable to make an adequate amount of antibodies to fight off a pathogen.

Treatment

Patients who suffer from CVID usually undergo antibiotic therapy paired with immunoglobulin therapy. As previously mentioned, Ms. Jones could not be treated for her low

IgA levels due to there being a high risk of her going into anaphylactic shock. However, Ms. Jones is currently being treated with a form of immunoglobulin therapy called, immunoglobulin infusion 10% subcutaneous (human with recombinant human hyaluronidase). The purpose of this treatment is to elevate her IgG levels helping her body fight off bacterial infections or other pathogenic attacks. The medicine is in the form of a liquid and is delivered underneath the skin (subcutaneously). The therapy is a lifelong treatment that is administered to the patient monthly or sometimes weekly. Currently, Ms. Jones is being treated weekly with 600 mg/kg of immunoglobulin infusion 10% subcutaneous averaging to about 54 grams a month.

There are many benefits that come with this immunoglobulin therapy. For Ms. Jones, a major benefit that comes with using immunoglobulin infusion 10% is that no IV infusion is needed. This means that for Ms. Jones, an elderly woman who most likely has damaged veins, her weekly treatment is much more tolerable as her veins are not subjected to an IV. Also, due to the fact that no IV is required, Ms. Jones's IgG therapy comes with a lot less systemic side effects that are usually experienced in patients who undergo a therapy that requires the penetration of a vein. This potentially high dosage and absorbance of the medication can lead to multiple side effects such as a low-grade fever, aching muscles/joints, and post-fusion headaches [5]. Also, when an IV is used to treat a CVID patient, there will be this "high phase" that is immediately followed by a "low phase" where the patient feels extremely well right after treatment but will then feel slow/tired a few days after treatment. These side effects are experienced much less by patients who take a subcutaneous route due to the fact that when the medication is administered subcutaneously, it is usually given at a lower dose and absorbed slowly. A side effect that can be experienced, however, can be a localized skin reaction that will fade with time [5].

Fun Facts

- Patients with CVID are usually not officially diagnosed until their third or fourth decade of life but may experience symptoms as early as childhood [3].
- It is estimated that CVID affects 1 in 25,000 to 1 in 50,000 people worldwide [2].
- In 90% of CVID cases, the causes are unknown but it is likely that the condition is developed by either genetic or environmental factors [2].

Multiple Choice Questions

1. What is/are possible autoimmune disorder(s) that a patient with CVID could develop?
   (a) Immune thrombocytopenia.
   (b) Autoimmune hemolytic anemia.
   (c) Rheumatoid arthritis.
   (d) All of the above

2. What is the reference range of IgG serum levels within a person's blood?
   (a) 645–1741 mg/dL
   (b) 66–433 mg/dL
   (c) 45–281 mg/dL
   (d) None of the above.

3. A person with CVID could produce an insufficient amount of ...
   (a) Red blood cells.
   (b) Immunoglobulins.
   (c) Platelets.
   (d) All of the above.

Answer
1. (d)
2. (a)
3. (b)

References

1. Common variable immune deficiency—Genetics Home Reference-NIH. https://ghr.nlm.nih.gov/condition/common-variable-immune-deficiency. Accessed 3 July 2020.
2. CVID Community Center. https://primaryimmune.org/about-primary-immunodeficiencies/specific-disease-types/common-variable-immune-deficiency. Accessed 3 July 2020.
3. Benefits of Hizentra. https://www.hizentra.com/benefits?source=Hizentra-PaidSearch-Competitor. Accessed 3 July 2020.
4. Immune Deficiency Foundation. https://primaryimmune.org/treatment-information/immunoglobulin-therapy. Accessed 3 July 2020.
5. Non-Hodgkin lymhoma early detection, diagnosis, and staging. https://www.cancer.org/cancer/non-hodgkin-lymphoma/detection-diagnosis-staging.html. Accessed 3 July 2020.

A 65-Year-Old Woman Presents with Chronic Cough: A Common Symptom with an Uncommon Cause

13

Robert A. Promisloff, Matthew Daidola, Mohammad Shigri, and Steven P. Kovacs

R. A. Promisloff
Drexel University College of Medicine, Philadelphia, PA, USA

M. Daidola (✉)
Touro College of Osteopathic Medicine, New York, NY, USA

M. Shigri
New York College of Podiatric Medicine, New York, NY, USA

S. P. Kovacs
Division of Pulmonary Medicine, UPMC Hamot Comprehensive Lung Center, UPMC Hamot Pulmonology, Erie, PA, USA
e-mail: kovacssp@upmc.edu

A. M. Szema (ed.), *Unusual Diseases with Common Symptoms*,
https://doi.org/10.1007/978-3-031-40280-7_13

cough
COUGH

Clinical Presentation

Mrs. G is a 60-year-old African-American lady who presented to our pulmonary office complaining of a chronic cough. Chronic cough is usually defined by a persistent cough for over 8 weeks. Surprisingly our patient Mrs. G had been coughing on-and-off for approximately 15 years. We learned that she had seen multiple doctors over that time for her symptoms and been prescribed an albuterol inhaler—which only helped her relentless coughing "a little." Recently, her condition had worsened to the point where she experienced shortness of breath with minimal physical exertion, like walking short distances or climbing a single flight of stairs. Mrs. G had no smoking or occupational history.

Mrs. G had a very extensive past medical history. She had diagnoses of high blood pressure, acid reflux disease, type II diabetes, and was recently diagnosed as having rheumatoid arthritis. Additionally, multiple doctors had told Mrs. G that she had both asthma and allergies. At one point, she even received allergy shots and again, the treatment only provided her "some relief." The arthritis drug Mrs. G was prescribed was known to have a suppressive impact on the immune system. In consideration she was given a tuberculosis skin test—which came back positive.

At present, Mrs. G was on an asthma inhaler with a bronchodilator and inhaled steroid, montelukast for asthma and allergies, amlodipine for high blood pressure, omeprazole for acid reflux, and non-sedating antihistamine. She also continued to occasionally use the albuterol inhaler. To further complicate matters, she had recently started on a daily dose of low dose of methotrexate and daily doses of prednisone. It appeared that none of her respiratory or allergy medicines alleviated the intensity or duration of her cough.

Her pulmonary workup included pulmonary function tests, which revealed mild restrictive disease. Simply put, her lungs were smaller than predicted but did not show an obstructive pattern, which we normally see in patients with asthma or tobacco-related lung disease. Allergic skin testing was positive for grass, trees, weeds, and dust mite antigens. A CT scan of her sinuses was unremarkable. A CT scan of the chest revealed multiple nodules or spots in both lungs that ranged from 2 mm to 10 mm in size. These nodules were randomly distributed throughout both lungs. A bronchoscopy was performed, which is an inspection of the airways with a video camera; this did not reveal any abnormalities or clues to the cause of her intractable coughing. Moreover, there was no evidence of infection or cancer.

Spirometry		Predicted	PRE	%Pred	POST	%Pred	%Change
FVC	(L)	2.36	1.17	50<	1.18	50<	0
FEV1	(L)	1.90	0.94	49<	1.08	57<	15
FEV1/FVC	(%)	80	80	99	92	115	15
FEV3	(L)	2.10	1.17	56<	1.18	56<	0
FEV3/FVC	(%)	89	100	112	100	112	0
FEF25-75%	(L/S)	2.28	0.74	33<	1.21	53<	62
FEFmax	(L/S)	5.47	6.61	121>	5.12	94	-22
FEF25%	(L/S)	5.12	6.61	129>	4.81	94	-27
FEF50%	(L/S)	3.04	1.24	41<	1.77	58<	43
FIF50%	(L/S)		3.00		0.97		-67
TET	(SEC)		2.88		3.17		10

Lung Volumes		Predicted	PRE	%Pred
TLC	(L)	3.87	2.74	71<
FRC	(L)	1.72	1.93	112
RV	(L)	1.48	1.68	113
VC	(L)	2.36	1.06	45<
IC	(L)	2.15	0.81	38<
ERV	(L)	0.23	0.25	107
RV/TLC	(%)	38	61	160>
He Equil.	(MIN)		2.75	

Diffusion	Predicted	PRE	%Pred
Dsb ml/min/mmHg	22.24	11.17	50<
Dsb(adj) ml/min/mmHg	22.24	11.17	50<
DCO(adj) ml/min/mmHg	22.24	12.07	54<
VA(rb) (L)	4.55	2.39	53<
VA(sb) (L)	4.55	2.21	49<
D/VA	4.89	5.05	103
D/VA(adj)	4.89	5.05	103
Volume Inspired (L)		1.09	
BHT (S)		10.68	

PT. WAS GIVEN TWO INHALATIONS OF PIRBUTEROL AS A BRONCHODILATOR.

(Highlight and put BOX around 50% and 49%) Pulmonary Function Study—Mrs. G

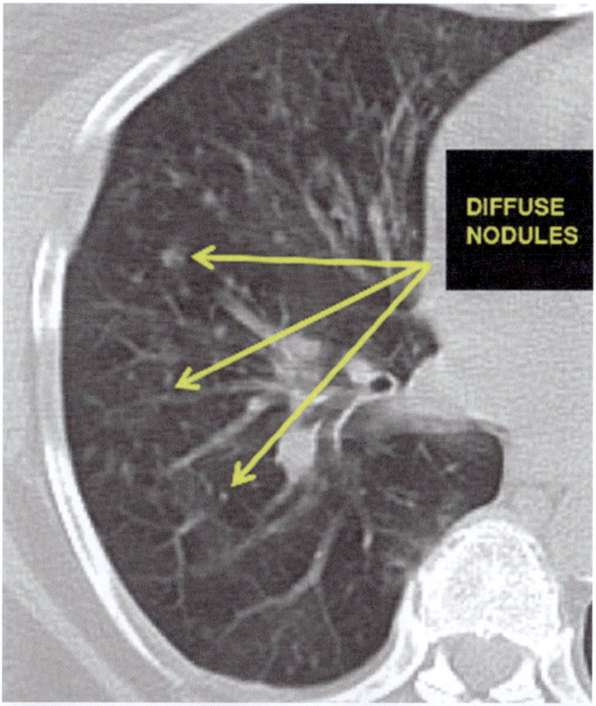

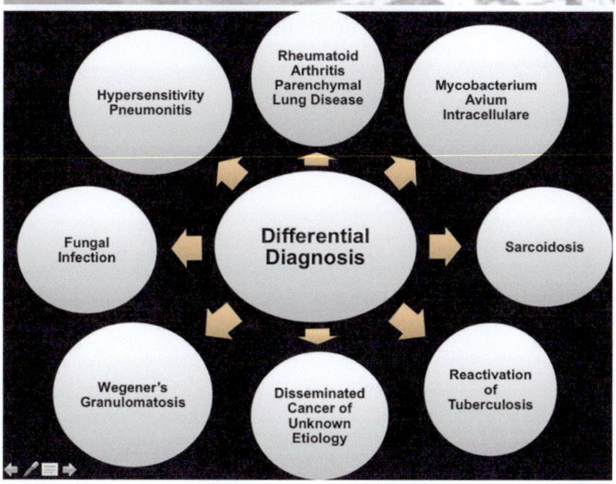

CT chest—Mrs. G

Diagnosis

After evaluation by the medical team and in discussion with the patient, we decided to perform a surgical lung biopsy. The etiology of the cough and multiple nodules in the lungs was incompletely understood. The differential diagnosis includes: a reactivation of tuberculosis, a complication from rheumatoid arthritis, sarcoidosis (which is more common in the African-American community), or cancer, whether primary or metastatic. Other types of infections, such as fungal infections, were also considered.

The patient tolerated the lung biopsy well. The result surprised the attending physicians. The initial pathology results were consistent with a neuroendocrine tumor and after the immunohistochemical stains (a common form of staining antigens with antibodies) were applied the final diagnosis reached was multiple carcinoid tumorlets. Among immunohistochemical stains that support the diagnosis of a carcinoid tumor are NSE, chromogranin, and synaptophysin. These tiny carcinoid tumors were the unexpected cause of the multiple nodules seen on the chest CT scan and the likely cause of her intractable coughing. Lung neuroendocrine tumors are thought to derive from neuroendocrine-producing cells throughout the body, most commonly found in the G.I. tract.

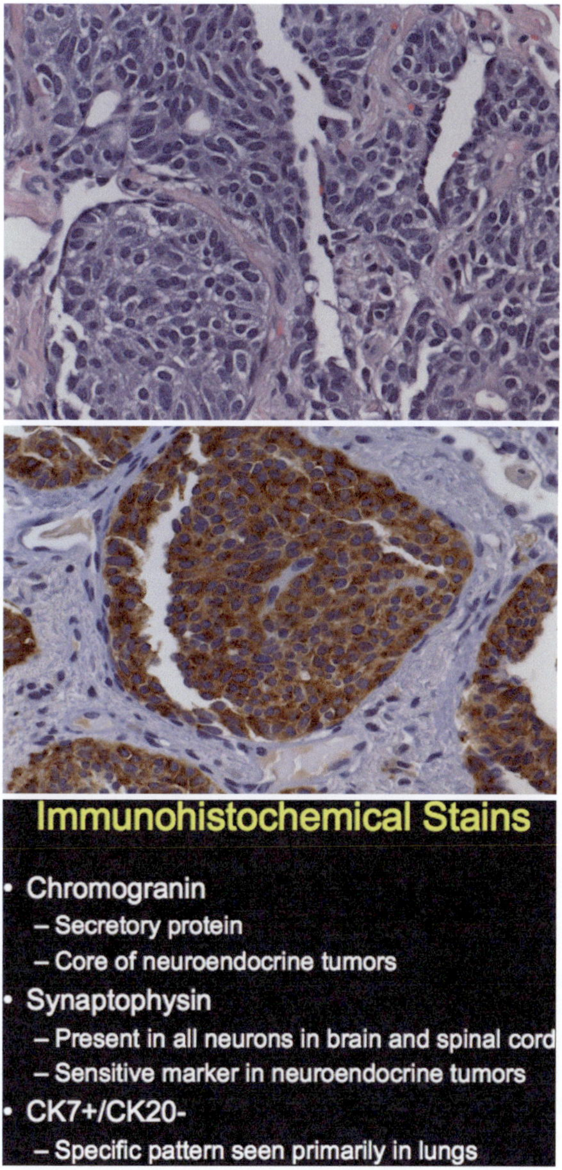

Surgical lung biopsy—Mrs. G

Discussion

Carcinoid tumorlets (tiny tumors) are a somewhat rare phenomenon. Commonly found in the G.I. tract, these neuroendocrine tumors can appear in the bronchial tree of the respiratory system. Oddly, in this case, they were scattered throughout the substance or parenchyma of the lung. It is noteworthy that these lung neuroendocrine tumors only account for 1–2% of all lung tumors in adults. They usually develop in patients 45–60 years old and the symptoms mainly consist of coughing and wheezing. Most are considered low-grade tumors. The association with smoking is not well established but in some studies up to two thirds of patients have been smokers. Carcinoids have a broad spectrum that range from slow-growing and benign to high grade, poorly differentiated, and malignant. Generalized proliferation of neuroendocrine cells is called DIPNECH, which stands for *diffuse idiopathic pulmonary neuroendocrine cell hyperplasia.* This proliferation would appear as multiple nodules or spots on an X-ray or CT scan of the lungs. Most patients with DIPNECH have respiratory symptoms, frequently presenting as a persistent cough.

Aside from coughing and shortness of breath, in rare circumstances, patients may develop *carcinoid syndrome*. Carcinoid syndrome is caused by the tumor releasing chemical substances. These tumor-released chemicals can cause flushing, diarrhea, and/or wheezing. Serotonin is a chemical commonly associated with the syndrome. Serotonin release can also cause valvular heart dysfunction. Lucky for Mrs. G, lung neuroendocrine tumors usually produce very little serotonin, as opposed to G.I. carcinoid tumors and account for a much lower rate of carcinoid syndrome.

So multiple carcinoid tumorlets or DIPNECH is a generalized proliferation of neuroendocrine cells that locally invade the lungs as multiple tiny tumors. Over time, these patients may also develop larger well-defined neuroendocrine tumors. Although coughing is a common clinical manifestation, this presentation is so unusual that it is rarely considered in the differential diagnosis of a chronic cough.

Treatment

Surgical resection is the best approach to treatment in patients who have lung carcinoids or neuroendocrine tumors that are localized, especially those producing carcinoid symptoms. In patients like ours, who have multiple tumors in both lungs, surgery is not feasible. One approach to this type of patient is a somatostatin analog, such as octreotide. This analog will bind to the tumor cell and inhibit the tumor from releasing the chemicals that cause the symptoms. Chemotherapy has been used but the responses are unpredictable. With carcinoid tumors in general, long-term survival is excellent.

Conclusion

Chronic cough is a very common symptom and is usually explained by common diagnoses like asthma, allergies, postnasal drip, or acid reflux disease. However, in this patient, the common diagnoses and treatment were woefully unsuccessful. The persistent symptoms (15 years!) broadened the possibilities beyond the scope of simple diagnostic measures. The diagnosis, in this case, would have been virtually impossible without a lung biopsy. There are instances when we, as physicians, must reach beyond our normal comfort level to diagnose entities that we have rarely or never seen in the past. In this instance a very common symptom was explained by a very rare disease.

Fun Facts

- Medical professionals consider a cough to be chronic only when its duration is over 8 weeks.
- Over 90% of chronic coughs will be diagnosed as either asthma, acid reflux disease, or postnasal drip.
- Sophisticated procedures such as bronchoscopy, CT scan, and lung biopsy are required in only a small number of patients who remain symptomatic and undiagnosed.
- When there are many spots or nodules on a chest X-ray or CT scan, many different entities need to be considered including

different types of infections, inflammatory diseases, and tumors.
- Chronic cough is the most common symptom that a patient will present to a physician and when the cough does not fit into the usual categories the physician may be forced to think "outside the box."

Multiple Choice Questions

1. A cough of 6 weeks duration is considered chronic.
   (a) True
   (b) False

2. A pulmonary function test can be very useful in diagnosing asthma as the cause of cough.
   (a) True
   (b) False

3. Acid reflux is a potential cause of chronic cough which worsens at night.
   (a) True
   (b) False

4. A CT scan of the chest is usually required to diagnose chronic cough.
   (a) True
   (b) False

5. Multiple spots or nodules on a chest CT could indicate infection, inflammation, or tumor.
   (a) True
   (b) False

Answers

1. (b)
2. (a)
3. (a)
4. (b)
5. (a)

14 Chronic Inflammatory Demyelinating Polyneuropathy (CIDP)

Niloofar Mirsaidi and Justin Kwan

He had been weightlifting when it happened. A cheerful, recently engaged 24-year-old Pennsylvanian, J. was at the gym attempting to do weighted squats with a staggering—yet for him normal—700-lb barbell rested on his shoulders, only to be met with surprising resistance. When he reduced the weight of the barbell to 200-lbs for bicep curls, he noticed again how he was unable to perform the exercise with his usual ease. Attempts at shoulder presses and latissimus dorsi muscle pull-downs were no different. Perhaps it was simply an off-day he thought to himself, unable to come up with any reasonable explanation for this new onset weakness. Uncertain and without answers, J. decided to go home to rest.

Within 1 week, J.'s tolerance for heavy weights decreased dramatically. In an unsettlingly short time, he found himself having difficulty rising up from a chair or the floor; his grip became weaker and he had multiple falls where his legs would give out or where he would trip over his toes. Routine daily activities, such as getting

N. Mirsaidi
Temple University, Lewis Katz School of Medicine, Philadelphia, PA, USA

J. Kwan (✉)
National Institute of Neurological Disorders and Stroke, National Institutes of Health, Bethesda, MD, USA
e-mail: justin.kwan@nih.gov

A. M. Szema (ed.), *Unusual Diseases with Common Symptoms*,
https://doi.org/10.1007/978-3-031-40280-7_14

dressed, putting on and taking off his socks and shoes, and drying himself with a towel, had become arduous tasks. J. also occasionally started to feel numbness and tingling in the middle digits of his hands. He additionally noticed muscle cramping in his hands and sporadic knee, elbow, and toe pain. Even those around him began to take note of certain changes, such as frequent dry coughs. He did not, however, experience any difficulty swallowing (dysphagia), drooping of his eyelids (ptosis), blurred vision, slurred speech (dysarthria), difficulty breathing (dyspnea) except on exertion, muscle pains (myalgias), or changes in his urine color—all of which may be attributed to a variety of neuromuscular diseases.

With no family or personal medical history of any musculoskeletal or neurologic illness and no history of medication use, J. grew increasingly concerned. He sought help from a neuromuscular specialist who had J. undergo extensive testing, including complete neurologic examination, blood work, cerebrospinal fluid analysis, and electromyography (or EMG, a test used to help identify nerve or muscle abnormalities). His physical examination was pertinent for the following:

- Mental status and cranial nerve exam: normal mental status, language, and cranial nerves
- Motor exam: diminished strength throughout his upper extremities (3/5 to 4-/5 on the Medical Research Council Scale), with greater weakness distally (i.e., toward his hands) and slight to moderately diminished strength in his lower extremities (4-/5 to 5/5)
- Reflexes: loss of deep tendon reflexes (areflexic) throughout
- Sensation: patchy areas of decreased sensation limited to foot and knee
- Gait: wide-based (indicative of the body's compensation for loss of balance)

Furthermore, electromyography (EMG) test and nerve conduction study revealed abnormalities indicating a demyelinating neuropathy. His bloodwork showed an elevated anti-SSA antibody titer (an antibody found in various autoimmune diseases) and analysis of his cerebrospinal fluid (CSF) revealed elevated protein with normal cell count.

Based on these findings, J. was diagnosed with chronic inflammatory demyelinating polyradiculoneuropathy (CIDP), a rare autoimmune disorder of the peripheral nervous system that requires long-term treatment. Fortunately, CIDP is a treatable disease in which therapeutic options include corticosteroids, intravenous immunoglobulins (IVIg), plasma exchange, as well as other immunosuppressive drugs. In the case of J., he was treated with IVIg for 5 days, which led to a drastic improvement of muscle strength (from 3/5 in all extremities to almost 5/5 by the end of his hospital stay). His initial treatment was complicated by headache and neck pain; however, this resolved upon initiation of a corticosteroid taper. J. was then discharged with home health nursing and treated with IVIg 0.5 g/kg daily for 2 days every 4 weeks with regular monitoring of his motor function, electrolytes, kidney function, and complete blood count (CBC). He again experienced mild headache as well as mild weakness after every initial infusion, but these would resolve the next day. Eventually, due to difficulties obtaining intravenous access by his home health nurse, he was switched to subcutaneous immunoglobulin (SCIg). He has since returned to his baseline state of health and resumed his normal weightlifting activities!

Discussion

CIDP is an acquired, immune-mediated demyelinating neuropathy (disease of the nerve). It is caused by abnormal cellular and humoral immune (immunity that involves antibody) responses against peripheral nerve antigens leading to demyelination of the nerves. Typical CIDP presents with remitting-relapsing or slowly progressive symmetric weakness of proximal and distal muscles, sensory symptoms (e.g., numbness or paresthesia) in a glove-and-stocking distribution, as shown in Image 14.1, and reduced or absent deep tendon reflexes [1]. CIDP symptoms typically progress over a span of 8 weeks or longer (making it a chronic process) [2].

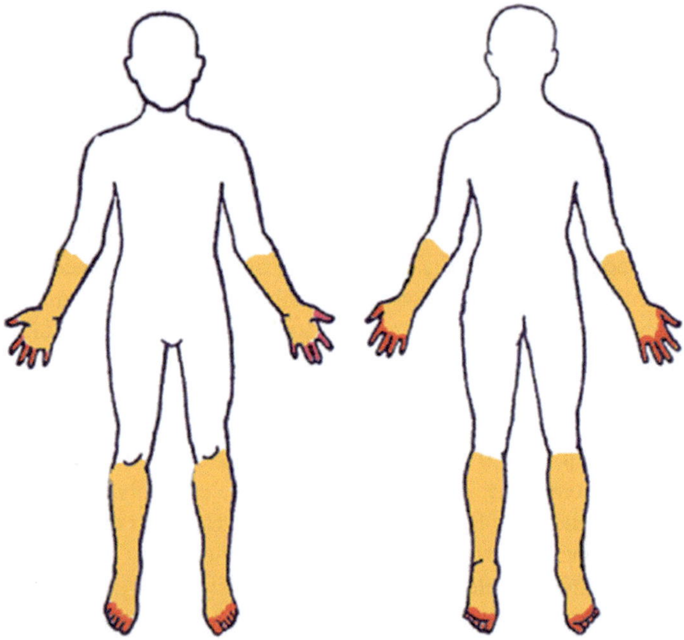

Image 14.1 Glove-and-stocking neuropathy

Rarely, CIDP can present acutely and may be indistinguishable from Guillain-Barré syndrome (GBS), a self-limiting demyelinating neuropathy [2]. It may only later be correctly diagnosed due to recurrence or prolonged symptom progression as well as associated events [1]. GBS typically does not recur and causes maximum weakness within 4 weeks, often following an event such as a viral infection or very rarely vaccination. Cranial and respiratory muscles and the autonomic nervous system are frequently affected in GBS [1]. CIDP, however, has a slower progression and is not associated with antecedent illness or vaccination. The term "subacute inflammatory demyelinating polyneuropathy" (SIDP) is used by some neuromuscular specialists to describe a rare form of demyelinating neuropathy that reaches its clinical nadir between 4 and 8 weeks [3, 4]. Patients with SIDP have overlapping features of GBS and CIDP.

In typical CIDP, the pattern of weakness is non-length-dependent (i.e., affecting both proximal and distal muscles to a similar degree). In fact, this phenotypic presentation is useful in diagnosing and thus treating patients, as it suggests an acquired demyelinating polyneuropathy. In contrast, patients with exclusively distal weakness have a broader differential diagnosis (including hereditary, metabolic, toxic, and vasculitic disorders) and the cause may not be identified until nerve conduction studies and laboratory workup can be performed [5]. Furthermore, variant CIDP patients with purely distal weakness and sensory loss may differ in both their clinical course and treatment response and are less likely to be correctly diagnosed and treated [5].

Several CIDP variants exist that differ in their degree of symmetry and motor or sensory involvement as well as their response to typical therapies. These include, but are not limited to, distal acquired demyelinating symmetric (DADS) neuropathy, multifocal acquired demyelinating sensory and motor (MADSAM) neuropathy (also known as Lewis-Sumner syndrome), pure motor or pure sensory CIDP, and focal CIDP [6].

As described in the patient case, the diagnosis of CIDP relies on typical clinical features, CSF findings, and electrophysiologic studies, as there is no biomarker available that is specific to this disease [7]. Nerve conduction study findings such as conduction block, prolonged distal latency, and slowed conduction velocity are features of a demyelinating neuropathy [1]. The pattern of nerve conduction abnormalities can help to determine the cause of the demyelinating neuropathy (i.e., patchy or multifocal in CIDP vs. more uniform in Charcot-Marie-Tooth disease) [1]. An EMG may also reveal secondary axonal loss [1]. Lumbar puncture shows elevated CSF protein with normal cell count (i.e., albuminocytologic dissociation) in CIDP patients. Pleocytosis (increase cell count), however, would suggest an alternate diagnosis [1]. While CSF studies are useful in supporting a CIDP diagnosis, they are not necessary in a patient with typical clinical features and electrophysiologic findings. Other clinical findings supportive of the CIDP diagnosis include MRI showing nerve root hypertrophy or enhancement with contrast, ultrasound showing nerve enlargement, nerve biopsy showing demyelinating features, and objective improvement after a trial of immunotherapy [1, 8].

Initial misdiagnosis, especially in atypical presentations, is not uncommon as CIDP shares features with other nerve and muscle disorders. These include GBS and Charcot-Marie-Tooth disease, as mentioned above, as well as motor neuropathies, spinal muscular atrophy, neuralgic amyotrophy [9], carpal tunnel syndrome, idiopathic sensory neuropathy, diabetic neuropathy, amyloidosis, and numerous other conditions [10]. There may be diagnostic confusion or error when the findings of the EMG/NCS and/or CSF are improperly contextualized or interpreted [7]. Of note, CIDP may occur in the setting of other disorders, such as Charcot-Marie-Tooth disease or hepatitis C [5].

The first-line treatment options for CIDP are corticosteroids, IVIg [1, 11], and plasmapheresis (i.e., therapeutic plasma exchange), each with its own advantages and disadvantages [5, 11–13]. When used in the short-term, corticosteroids are efficacious and generally inexpensive with an associated longer remission period [11]; however, long-term use of high dose corticosteroids can cause multiple side effects (e.g., steroid-induced osteoporosis and myopathy, osteonecrosis, weight gain, increased risk for infection, suppressed adrenal gland function, cataracts, cardiovascular effects, and neuropsychiatric effects [14]) and may be poorly tolerated, especially in older individuals with other medical conditions [15]. Plasma exchange shows short-term benefit with patients responding within 4 weeks of treatment but is rarely used as a monotherapy due to a high percentage of patients deteriorating weeks to months following treatment [14, 15]. Plasma exchange is not widely used due to the cost and labor intensive nature of the therapy as well as the need for large bore peripheral venous access or arteriovenous fistula placement [14, 15]. IVIg is a frequently prescribed therapy with a 50–70% response rate and has been the standard of care for over two decades due to its rapid onset of action and more sustained response when compared to plasma exchange [11]. IVIg therapy can cause a range of adverse effects from mild symptoms such as headache and fatigue, to more serious side effects such as aseptic meningitis, congestive heart failure, and renal failure [15]. The financial burden for patients is also substantial, with a mean cost near $120,000 in those treated with IVIg alone and greater than $133,000 in those receiving IVIg in addition to another CIDP therapy over a 2-year period [11]. Other immunosuppressive therapies, such as azathioprine, cyclophosphamide, methotrexate and rituximab may also be used in CIDP but are typically reserved for patients who fail or inadequately respond to corticosteroid and IVIg therapy [5, 15].

It is important to note that response to individual treatments can vary and patients with CIDP variants may have a less predictable response to these therapies [10].

This article was partially prepared while Dr. Kwan was employed at Temple University. The opinions expressed in this article are the author's own and do not reflect the view of the National Institutes of Health, the Department of Health and Human Services, or the United States Government.

References

1. Reynolds J, Sachs G, Stavros K. Chronic inflammatory demyelinating polyradiculoneuropathy (CIDP): clinical features, diagnosis, and current treatment strategies. R I Med J (2013). 2016;99(12):32–5.
2. Ripellino P, Fleetwood T, Cantello R, Comi C. Treatment of chronic inflammatory demyelinating polyneuropathy: from molecular bases to practical considerations. Autoimmune Dis. 2014;2014:201657. https://doi.org/10.1155/2014/201657.
3. Hughes R, Sanders E, Hall S, Atkinson P, Colchester A, Payan P. Subacute idiopathic demyelinating polyradiculoneuropathy. Arch Neurol. 1992;49(6):612–6. https://doi.org/10.1001/archneur.1992.00530300044009. PMID: 1317700.
4. Oh SJ, Kurokawa K, de Almeida DF, Ryan HF Jr, Claussen GC. Subacute inflammatory demyelinating polyneuropathy. Neurology. 2003;61(11):1507–12. https://doi.org/10.1212/01.wnl.0000096166.28131.4c. PMID: 14663033.
5. Saperstein DS, Katz JS, Amato AA, Barohn RJ. Clinical spectrum of chronic acquired demyelinating polyneuropathies. Muscle Nerve. 2001; 24(3):311–24.https://doi.org/10.1002/1097-4598(200103)24:3<311::aid-mus1001>3.0.co;2-a.
6. Doneddu PE, Cocito D, Manganelli F, on behalf of the Italian CIDP Database study group, et al. Atypical CIDPiagnostic criteria, progression and

treatment response. Data from the Italian CIDP database. J Neurol Neurosurg Psychiatry. 2019;90:125–32.

7. Allen J, Lewis R. CIDP diagnostic pitfalls and perception of treatment benefit. Neurology. 2015;85(6):498–504. https://doi.org/10.1212/WNL.0000000000001833.
8. Van den Bergh PY, Hadden RD, Bouche P, Cornblath DR, Hahn A, Illa I, Koski CL, Léger JM, Nobile-Orazio E, Pollard J, Sommer C, van Doorn PA, van Schaik IN, European Federation of Neurological Societies; Peripheral Nerve Society. European Federation of Neurological Societies/Peripheral Nerve Society guideline on management of chronic inflammatory demyelinating polyradiculoneuropathy: report of a joint task force of the European Federation of Neurological Societies and the Peripheral Nerve Society - first revision. Eur J Neurol. 2010;17(3):356–63. https://doi.org/10.1111/j.1468-1331.2009.02930.x. Erratum in: Eur J Neurol. 2011 May;18(5):796. PMID: 20456730.
9. Bunschoten C, Jacobs BC, Van den Bergh PYK, Cornblath DR, van Doorn PA. Progress in diagnosis and treatment of chronic inflammatory demyelinating polyradiculoneuropathy. Lancet Neurol. 2019;18(8):784–94. https://doi.org/10.1016/S1474-4422(19)30144-9. Epub 2019 May 7. PMID: 31076244.
10. Neligan A, Reilly MM, Lunn MP. CIDP: mimics and chameleons. Pract Neurol. 2014;14(6):399–408. https://doi.org/10.1136/practneurol-2014-000831. Epub 2014 Jul 17. PMID: 25035142.
11. Guptill JT, Runken MC, Eaddy M, Lunacsek O, Fuldeore RM. Treatment patterns and costs of chronic inflammatory demyelinating polyneuropathy: a claims database analysis. Am Health Drug Benefits. 2019;12(3):127–35.
12. Dyck PJ, Daube J, O'Brien P, et al. Plasma exchange in chronic inflammatory demyelinating polyradiculoneuropathy. N Engl J Med. 1986;314(8):461–5. https://doi.org/10.1056/NEJM198602203140801.
13. Hahn AF, Bolton CF, Pillay N, et al. Plasma-exchange therapy in chronic inflammatory demyelinating polyneuropathy. A double-blind, sham-controlled, cross-over study. Brain. 1996;119(Pt 4):1055–66. https://doi.org/10.1093/brain/119.4.1055.
14. Yasir M, Goyal A, Bansal P, et al. Corticosteroid adverse effects. [Updated 2020 July 4]. In: StatPearls [Internet]. Treasure Island, FL: StatPearls Publishing; 2020. Available from: https://www.ncbi.nlm.nih.gov/books/NBK531462/.
15. Gorson KC. An update on the management of chronic inflammatory demyelinating polyneuropathy. Ther Adv Neurol Disord. 2012;5(6):359–73. https://doi.org/10.1177/1756285612457215.

Chronic Scratching and Neurodermatitis Hatching

15

Gregory T. Smith

Vignette
Have you ever thought about being itchy and then started scratching at your skin? Charlotte Hass is a 67-year-old woman who presented with dry skin, hyperpigmentation, and general discoloration on her upper back. As she grew more itchy, she would scratch at her skin, causing a bothersome itch-scratch cycle [1]. Her back would inflame, become scaly, and ultimately exacerbate the itch-scratch cycle. But what exactly is happening in the body during this condition?

Charlotte's past medical history included renal cell carcinoma, stress-induced hair loss, multinodular goiter, chronic pain, and dry skin. The renal cell carcinoma was treated with a subsequent nephrectomy. Additional past surgical history includes cataract removal. Charlotte's family history includes a sister with bilateral breast cancer who is a carrier for the mutant *BRCA* gene.

Background/Salient Features of Case

Lichen simplex chronicus, or neurodermatitis, refers to a hypertrophied epidermal layer resulting from repeated contact with the skin due to rubbing or scratching possibly triggered by

G. T. Smith (✉)
Department of Pharmacological Sciences, Stony Brook University, Stony Brook, NY, USA

A. M. Szema (ed.), *Unusual Diseases with Common Symptoms*,
https://doi.org/10.1007/978-3-031-40280-7_15

psychological stressors [1]. This pruritic dermatitis occurs in approximately 12% of the population [1]. According to Charifa et al. (2020), histopathology of the infected area may reveal hyperkeratotic plaques, pseudoepitheliomatous hyperplasia, papillary dermal fibrosis, mild spongiosis, perivascular and interstitial inflammatory histiocytes, acanthosis, a prominent granular cell layer, and occasional eosinophils in the dermis [1].

But what *exactly* is causing the itchiness and the immune response? At the moment a person feels itchy, there is a complex interplay between the epithelium, the peripheral nervous system, spinal cord, immune system, and central nervous system [2]. In the case of Charlotte, scratching can activate the epithelial stress response, leading to the release of inflammatory cytokines, such as thymic stromal lymphopoietin (TSLP) and interleukin-33 (IL-33) [2]. Once released, these cytokines can bind to and activate pruriceptors, or the neurons involved in the itchy sensation [2]. Epithelial stress caused by chronic itching can also lead to the release of proteins such as kallikreins (KLK) and cathepsin S, which also activate pruriceptors [2]. Conversely, an inflammatory response independent of the epithelial stress can also trigger the itchy sensation [2]. Immune cells such as mast cells, basophils, Th2 cells, and ILC2 cells can further worsen the itch upon activation of pruriceptors through the release of tryptase, histamine, IL-31, IL-4, and IL-13 [2].

In the case of the patient, psychological stress may have promoted the itch feeling through activation of the local pruriceptive neurons, possibly through neuroimmunological pathways. Once the itch-scratch cycle began, the chronic rubbing and scratching could have triggered the subsequent lichenification, or thickening of the skin. Together, this condition spans an axis that involves interplay between the central and peripheral nervous systems, integumentary system, and immune system [2].

Diagnosis

Charlotte was diagnosed with lichen simplex chronicus, otherwise known as neurodermatitis, as well as leukocytoclastic vasculitis. In the image below, the lesions are hyperpigmented and thickened. The skin also contains prominent skin lines as well as the presence of multiple small papules. Collectively, the process of developing these clinical phenotypes is known as lichenification [3] (Fig. 15.1).

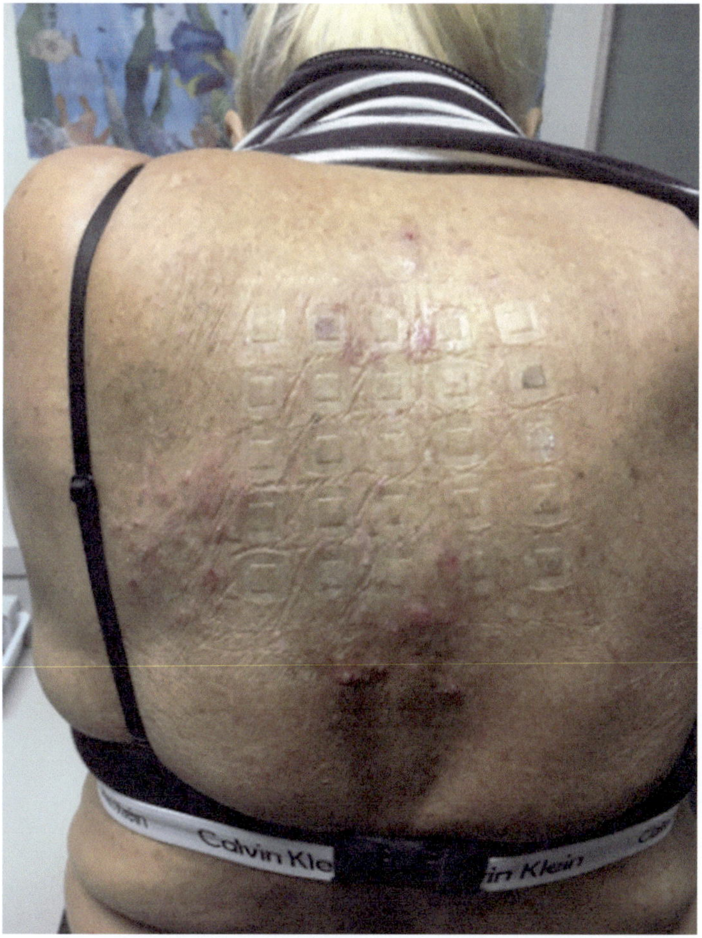

Fig. 15.1 Image of the infected area

Treatment

60 g 2% Eucrisa® ointment, or crisaborole, was prescribed to treat the pruritic dermatitis by applying to the affected area twice per day. The mechanism of action is through inhibiting phosphodiesterase-4, or PDE-4, an enzyme that is highly expressed in immune cells [4, 5]. Phosphodiesterase inhibitors ultimately result in an increase of intracellular cAMP, leading to varying downstream effects depending on the target tissue and cell type [5]. With respect to dermatological pharmacodynamics, the increase in cAMP leads to inhibition of pro-inflammatory mediators such as IL-6, IL-12, and TNF-α [5]. Thus, crisaborole works through an anti-inflammatory pathway by treating immune-mediated dermatological diseases [5].

Fun Facts

1. Pruriceptive neurons are implicated in the itch-scratch cycle which involves immunological, neurological, and dermatological components [2].
2. Psychological stress may manifest as a dermatological condition [1].
3. Lichenification involves thickening of the skin which can be triggered by constant rubbing [3].

Multiple Choice Questions

1. Which of the following cellular mediators are NOT implicated in the epithelial stress response involved in pruriceptive activation?
   (a) Thymic stromal lymphopoietin (TSLP)
   (b) Cathepsin S
   (c) IL-33
   (d) IL-13

2. The mechanism of crisaborole pharmacodynamics involves:
   (a) Normalization of keratinocyte differentiation
   (b) Agonistic binding to phosphodiesterase-2
   (c) Inhibition of phosphodiesterase-4
   (d) Normalization of follicular epithelial differentiation

3. Epithelial stress from chronic itching can lead to the release of kallikreins (KLK) and cathepsin S which lead to activation of:
   (a) Interoreceptors
   (b) Exteroreceptors
   (c) Nociceptors
   (d) Pruriceptors

Answers
1. (d)
2. (c)
3. (d)

References

1. Charifa A, Badri T. Lichen simplex chronicus. In: StatPearls [Internet]. StatPearls Publishing; 2020.
2. Mack MR, Kim BS. The itch–scratch cycle: a neuroimmune perspective. Trends Immunol. 2018;39(12):980–91.
3. Aboobacker S, Limaiem F. Lichenification. 2019.
4. https://www.fda.gov/media/102760/download#:~:text=3%20What%20are%20the%20proposed,adenosine%20monophosphate%20(cAMP)%20levels.
5. Moustafa F, Feldman SR. A review of phosphodiesterase-inhibition and the potential role for phosphodiesterase 4-inhibitors in clinical dermatology. Dermatol Online J. 2014;20(5):22608.

Index

A. M. Szema (ed.), *Unusual Diseases with Common Symptoms*, https://doi.org/10.1007/978-3-031-40280-7

MIX
Papier aus verantwortungsvollen Quellen
Paper from responsible sources
FSC® C105338

If you have any concerns about our products,
you can contact us on
ProductSafety@springernature.com

In case Publisher is established outside the EU,
the EU authorized representative is:
Springer Nature Customer Service Center GmbH
Europaplatz 3, 69115 Heidelberg, Germany

Printed by Libri Plureos GmbH
in Hamburg, Germany